Manuel F.M. Barros, Jorge M.C. Guilherme and Nuno C.G. Horta

Analog Circuits and Systems Optimization Based on Evolutionary Computation Techniques

Studies in Computational Intelligence, Volume 294

Editor-in-Chief
Prof. Janusz Kacprzyk
Systems Research Institute
Polish Academy of Sciences
ul. Newelska 6
01-447 Warsaw
Poland
E-mail: kacprzyk@ibspan.waw.pl

Further volumes of this series can be found on our homepage: springer.com

Vol. 268. Johann M.Ph. Schumann and Yan Liu (Eds.)
Applications of Neural Networks in High Assurance Systems, 2009
ISBN 978-3-642-10689-7

Vol. 269. Francisco Fernández de de Vega and Erick Cantú-Paz (Eds.)
Parallel and Distributed Computational Intelligence, 2009
ISBN 978-3-642-10674-3

Vol. 270. Zong Woo Geem
Recent Advances In Harmony Search Algorithm, 2009
ISBN 978-3-642-04316-1

Vol. 271. Janusz Kacprzyk, Frederick E. Petry, and Adnan Yazici (Eds.)
Uncertainty Approaches for Spatial Data Modeling and Processing, 2009
ISBN 978-3-642-10662-0

Vol. 272. Carlos A. Coello Coello, Clarisse Dhaenens, and Laetitia Jourdan (Eds.)
Advances in Multi-Objective Nature Inspired Computing, 2009
ISBN 978-3-642-11217-1

Vol. 273. Fatos Xhafa, Santi Caballé, Ajith Abraham, Thanasis Daradoumis, and Angel Alejandro Juan Perez (Eds.)
Computational Intelligence for Technology Enhanced Learning, 2010
ISBN 978-3-642-11223-2

Vol. 274. Zbigniew W. Raś and Alicja Wieczorkowska (Eds.)
Advances in Music Information Retrieval, 2010
ISBN 978-3-642-11673-5

Vol. 275. Dilip Kumar Pratihar and Lakhmi C. Jain (Eds.)
Intelligent Autonomous Systems, 2010
ISBN 978-3-642-11675-9

Vol. 276. Jacek Mańdziuk
Knowledge-Free and Learning-Based Methods in Intelligent Game Playing, 2010
ISBN 978-3-642-11677-3

Vol. 277. Filippo Spagnolo and Benedetto Di Paola (Eds.)
European and Chinese Cognitive Styles and their Impact on Teaching Mathematics, 2010
ISBN 978-3-642-11679-7

Vol. 278. Radomir S. Stankovic and Jaakko Astola
From Boolean Logic to Switching Circuits and Automata, 2010
ISBN 978-3-642-11681-0

Vol. 279. Manolis Wallace, Ioannis E. Anagnostopoulos, Phivos Mylonas, and Maria Bielikova (Eds.)
Semantics in Adaptive and Personalized Services, 2010
ISBN 978-3-642-11683-4

Vol. 280. Chang Wen Chen, Zhu Li, and Shiguo Lian (Eds.)
Intelligent Multimedia Communication: Techniques and Applications, 2010
ISBN 978-3-642-11685-8

Vol. 281. Robert Babuska and Frans C.A. Groen (Eds.)
Interactive Collaborative Information Systems, 2010
ISBN 978-3-642-11687-2

Vol. 282. Husrev Taha Sencar, Sergio Velastin, Nikolaos Nikolaidis, and Shiguo Lian (Eds.)
Intelligent Multimedia Analysis for Security Applications, 2010
ISBN 978-3-642-11754-1

Vol. 283. Ngoc Thanh Nguyen, Radoslaw Katarzyniak, and Shyi-Ming Chen (Eds.)
Advances in Intelligent Information and Database Systems, 2010
ISBN 978-3-642-12089-3

Vol. 284. Juan R. González, David Alejandro Pelta, Carlos Cruz, Germán Terrazas, and Natalio Krasnogor (Eds.)
Nature Inspired Cooperative Strategies for Optimization (NICSO 2010), 2010
ISBN 978-3-642-12537-9

Vol. 285. Roberto Cipolla, Sebastiano Battiato, and Giovanni Maria Farinella (Eds.)
Computer Vision: Recognition, Registration and Reconstruction, 2010
ISBN xxx

Vol. 286. A. Bolshoy, Z. Volkovich, V. Kirzhner, and Z. Barzilay (Eds.)
Genome Clustering: From Linguistic Models tO Classification of Genetic Texts, 2010
ISBN xxx

Vol. 287. Dan Schonfeld, Caifeng Shan, Dacheng Tao, and Liang Wang (Eds.)
Video Search and Mining, 2010
ISBN xxx

Vol. 288-293. xxx

Vol. 294. Manuel F.M. Barros, Jorge M.C. Guilherme, and Nuno C.G. Horta
Analog Circuits and Systems Optimization based on Evolutionary Computation Techniques, 2010
ISBN 978-3-642-12345-0

Manuel F.M. Barros, Jorge M.C. Guilherme,
Nuno C.G. Horta

Analog Circuits and Systems Optimization Based on Evolutionary Computation Techniques

Manuel F.M. Barros
Instituto de Telecomunicações
Instituto Politécnico de Tomar
Av. Rovisco Pais 1
1049-001 Lisboa
Portugal
E-mail: fmbarros@ipt.pt

Nuno C.G. Horta
Instituto de Telecomunicações
Instituto Superior Técnico
Av. Rovisco Pais 1
1049-001 Lisboa
Portugal
E-mail: n.horta@ieee.org

Jorge M.C. Guilherme
Instituto de Telecomunicações
Instituto Politécnico de Tomar
Av. Rovisco Pais 1
1049-001 Lisboa
Portugal
E-mail: jorge.guilherme@ipt.pt

ISBN 978-3-642-26323-1 ISBN 978-3-642-12346-7(eBook)

DOI 10.1007/978-3-642-12346-7

Studies in Computational Intelligence ISSN 1860-949X

Softcover reprint of the hardcover 1st edition 2010

Typeset & Cover Design: Scientific Publishing Services Pvt. Ltd., Chennai, India.

Printed on acid-free paper

9 8 7 6 5 4 3 2 1

springer.com

Manuel F.M. Barros
To Fatuxa, Catarina, Lucas and Joaquim

Jorge M.C. Guilherme
To Paula, Patricia and Inês

Nuno C. G. Horta
To Carla, João and Tiago

Preface

The microelectronics market, with special emphasis to the production of complex mixed-signal systems-on-chip (SoC), is driven by three main dynamics, time-to-market, productivity and managing complexity. Pushed by the progress in nanometer technology, the design teams are facing a curve of complexity that grows exponentially, thereby slowing down the productivity design rate. Analog design automation tools are not developing at the same pace of technology, once custom design, characterized by decisions taken at each step of the analog design flow, relies most of the time on designer knowledge and expertise. Actually, the use of design management platforms, like the Cadences Virtuoso platform, with a set of integrated CAD tools and database facilities to deal with the design transformations from the system level to the physical implementation, can significantly speed-up the design process and enhance the productivity of analog/mixed-signal integrated circuit (IC) design teams. These design management platforms are a valuable help in analog IC design but they are still far behind the development stage of design automation tools already available for digital design. Therefore, the development of new CAD tools and design methodologies for analog and mixed-signal ICs is essential to increase the designer's productivity and reduce design productivitygap.

The work presented in this book describes a new design automation approach to the problem of sizing analog ICs. The developed design optimization tool, GENOM, is based on a modified genetic algorithm (GA) kernel and incorporates heuristic knowledge on the control mechanism allowing a significant reduction on the required number of generations and, therefore, iterations to reach the optimal solution. However, the optimization process, employing a simulation-based approach with a kernel based on stochastic optimization techniques is clearly a computational intensive task typified by high dimension search spaces and high cost function evaluations. A step forward to enhance the efficiency of the implemented optimization tool corresponds to the introduction of behavior modeling techniques. The model introduced in this paper follows a supervised learning strategy based on support vector machines (SVM) which, together with an evolutionary strategy, is used to create feasibility models in order to efficiently prune the design search space during the optimization process, thus, reducing the overall number of required evaluations.

The book is organized in seven chapters. The first one, the introduction, presents the motivation and outlines the original goals for this research work.

Chapter 1 provides an overview of the thesis motivations, research goals and main contributions.

Chapter 2 presents a state-of-the-art review in analog IC design automation field by analyzing and comparing methods, strategies and tools presented in literature, including some commercial tools.

Chapter 3 starts with an overview on computation techniques to solve nonlinear optimization problems with focus on evolutionary optimization algorithms. Then, it introduces a new optimization kernel based on genetic algorithms applied to analog circuit optimization. It includes a detailed description of the fitness function, the genetic operators and design methodology in order to obtain an efficient and robust analog circuit design.

Chapter 4 explores the main learning techniques used to manage large amount of information, to discover complex relationships among various factors and extract meaningful knowledge to improve the efficiency and quality of decision-making. In particular, it discusses the use and the integration of a learning model based in support vector machine (SVM) in order to improve the evolutionary optimization strategy for analog circuit design applications introduced in chapter 3.

Chapter 5 describes the analog design environment and architecture of GENOM optimization tool. It discusses the methodology, representation and architecture issues, giving details of the analog IC design representation, interfaces between the synthesizer and evaluation algorithms, and software architecture. The main options taken in this work approach will be described and justified.

Chapter 6 presents several synthesis experiments, demonstrating the capabilities of the system and providing some insight into factors that affect the synthesis process. The suite of test circuits is taken from standard text books and technical papers. The first section describes the performance metrics, the algorithm optimization strategy and input data of each the experiment. The resulting performances computed automatically by the optimization tool during the evolutionary process are delivered to the user in the form of output reports or by dynamic graphics or reports. Apart from accuracy, mean and standard deviation of execution time and evaluation cycles are also presented. Additionally, information regarding the circuit, such as circuit sizing, corner information and performance are also specified.

Finally, chapter 7 presents the conclusions and the research contributions of the thesis and the improvements that are possible to GENOM.

Manuel F.M. Barros
Jorge M.C. Guilherme
Nuno C.G. Horta

Contents

List of Figures

List of Tables

List of Abbreviations

AI	Artificial Intelligence
AMD	Advanced Micro Devices
AMS	Austrian Micro Systems
ADC	Analog-to-Digital Converter
ADSL	Asymmetrical Digital Subscriber Line
AIDA	Analog Integrated Circuit Design Automation platform
ASIC	Application Specific Integrated Circuits
BiCMOS	Bipolar Complementary Metal-Oxide Semiconductor
CAD	Computer Aided Design
CMFB	Common Mode Feedback Amplifier
CMOS	Complementary Metal Oxide Semiconductor
CMR	Common-Mode Range
CMRR	Common-Mode Rejection Ratio
DA	Design Automation
DAC	Digital-to-Analog Converter
DSP	Digital Signal Processing
EDA	Electronic Design Automation
EP	Evolutionary Programming
EA	Evolutionary Algorithms
GA	Genetic Algorithms
GBW	Gain-Bandwidth Product
GP	Geometrical Programming
MOO	Multi-Objective Optimization
NMOS	N-channel MOSFET
NN	Neural Networks
OPAMPS	Operational amplifiers
OR	Output Range
OTA	Operational Transconductance Amplifier
PMOS	P-channel MOSFET
PSRR	Power Supply Rejection Ratio
SNR	Signal-to-Noise Ratio
GUI	Graphical User Interface
IC	Integrated Circuit
IP	Intellectual Property
OR	Output Range

RF	Radio Frequency
SA	Simulated Annealing
SoC	System-on-Chip
SNR	Signal-to-Noise-Ratio
SVM	Support Vector Machines
SPICE	Simulated Program with Integrated Circuits Emphasis
VLSI	Very Large Scale Integration
THD	Total Harmonic Distortion
UGBW	Unity-Gain Bandwidth
VHDL	Very High Speed Integrated Circuits Hardware Description Language
VHDL-AMS	Analog Mixed Signal VHDL

1 Introduction

Abstract. This chapter presents the motivation to the research work in the area of analog integrated circuit (IC) design automation, i.e., outlines the market and technological evolution, characterizes the analog IC design, discusses the available CAD solutions and, finally, describes goals for the this work.

1.1 Microelectronics Market and Technology Evolution

The microelectronics market trends present an ever-increasing level of integration with special emphasis on the production of complex mixed-signal systems-on-chip (SoC), as a consequence of the boom in telecom devices, wireless communications, electronic consumer products, etc. These devices integrate complex digital cores with analog and RF functions on a single chip [1]-[2]. Fig. 1.1 illustrates the relevance of analog circuitry in this renewed invigorating market showing the evolution of SoC percentage that will contain analog parts. According to IBS Corporation, digital/mixed-signal SoCs accounted for approximately 20% of worldwide SoCs in 2001[3]. The tendency curve in the left graph shows that the percentage continues to rise, and it was around 75% in 2006. Driven by the demand to provide higher performances, i.e., increasing functionalities with less power consumption, semiconductor manufacturers developed newer technologies allowing an exponential increase in IC density, described by the well known *Moore's* law.

The famous Moore's law observed for the first time in 1965, which states that transistor density on integrated circuits doubles about every two years, is still applied nowadays. It is a measure of the technological progress verified in semiconductor

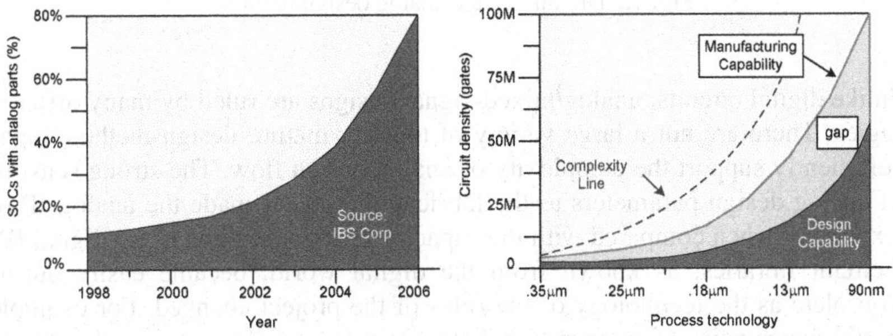

Fig. 1.1 (a) SoCs including analog circuitry by year [3], (b) Digital circuit density by process technology

M.F.M. Barros et al.: Analog Circuits and Systems Optimization, SCI 294, pp. 1–18.
springerlink.com

fabrication technologies but now, more than ever it has become a source of concerns. As far as new process technologies evolve, more and more functionalities can be integrated, increasing as a side effect, the complexity of IC design. A large augment of parasitics and leakage currents that make chips fail are the first visible sign. Actually, while the exponential growth in capacity moves on into the nanometer technology domain, the design teams are facing a curve of complexity that rises exponentially faster, thereby slowing down the productivity design rate. This phenomenon, referred as the design productivity gap [3] is illustrated in Fig. 1.1 (right) and represents the growing difference between the improvement in manufacturing productivity needed to satisfy the pressures of the market and the progress in productivity achieved by Computer-Aided-Design (CAD) development teams. The development of new CAD tools and design methodologies is essential to increase the designer's productivity and reduce design productivity gap. However, the development of mixed-signal SoC designs can be very challenging since the design and verification processes for both digital and analog sections are supported by design automation tools in different stages of maturity. The analog section, despite of typically occupying a minor fraction of the overall circuit area (20% approximately as illustrated in Fig. 1.2), is the bottleneck in terms of design time by being far more complex than the digital counterpart. Generally, digital design is based on well-established practices supported by well-defined automated synthesis methodologies and tools. As a result, digital intellectual property (IP) reuse is a common practice available through different CAD companies leading to an increase in the design productivity.

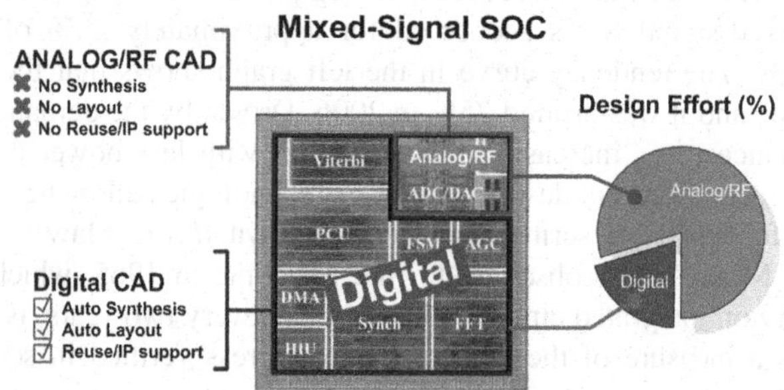

Fig. 1.2 Digital versus Analog design reality

Unlike digital circuits, analog/mixed-signal designs are ruled by many different strategies. There are not a large variety of tools or mature design methodologies that efficiently support the complexity of analog design flow. The strong sensitivity of analog design parameters to the fabrication process made the analog IP reuse expensive when compared with the capacity of acquiring and using digital IPs. The circuit libraries, as known from the digital world, became easily out-of-date/obsolete as the technology or the rules of the project changed. For example, the performance noise of an operational amplifier degenerates with smaller dimension technologies whereas the gain DC of small signal improves. This way, the upgrade of new capabilities and functionalities offered by a recent technology became more difficult.

Finally, despite the great progress achieved in the past few years concerning analog circuit technologies, analog design tools and methods are still far from reaching a mature stage.

1.2 Analog Integrated Circuit Design

1.2.1 Analog Design Issues

The difficulty of analog circuit design is higher than digital circuit design. The signals of the digital technology are more tolerant to noise levels, by having only two possible values each one with a large tolerance range. The Boolean functionality of digital systems represented through the use of high level description languages and mature digital CAD tools facilitates the automation of design tasks. Analog design, on the other hand, deals with an infinite or continuous range of values, which force it to exploit the physics of the fabrication process to achieve high performance designs. Therefore, second-order and third-order effects that are not so critical for digital design become a major problem for analog designs [1][4].

Due to these problems, the employment of standard cell libraries for analog design is not widespread since each analog cell is characterized by several continuous parameters (including power dissipation, DC gain, bandwidth, phase mar-gin, slew rate, noise, power, area, etc.) which produce hundreds or thousands of instances for each cell with different performance measures [1]. Additionally, some of them will not create functional solutions. Hence, the technology migration and the retargeting of analog designs usually require substantial redesign of the circuit, unlike the digital design [4].

Today, one of the major challenges faced by semiconductor companies is how to increase yield of their circuits. During the layout and production phase, the appearance of parasitic effects, device mismatch and changes in environment conditions have a negative influence on the behavior of the designed circuit, changing the performance parameters and leading to undesired performance of the circuit. One way to minimize the effect of these variations consists in the use of evaluation and compensation techniques during the design phase, combined with a careful layout [5]. Corners analysis in extreme variation points is an example of such techniques consisting in the circuit simulation for different operational conditions, for example, different temperatures and process variations.

The above described problems deal with the characteristic of analog circuits, which makes the analog design a hard task. The performance of analog blocks is a key factor in the success of an integrated circuit. The time-to-market imposes a first time right on both digital and analog blocks in integrated circuits and systems on chip. To create a complete solution and achieve the target of first time right in analog systems, careful methodologies, tool flows and an appropriate set of tools must be used by analog designers. These techniques and tools will be summarized in the next sections.

1.2.2 The Hierarchical Decomposition Model

The increase in complexity of analog and mixed-signal integrated circuits leads to the general use of IC design methodologies based on divide-to-conquer strategies, sustained by a hierarchical decomposition model that define the top-down design and bottom-up verification flows [4][6][7][8][9][10] and a set of design tasks in each hierarchical level [4][6][7][8]. In order to illustrate the above concepts of a typical design flow, Fig. 1.3 shows the design of an Analog Front-end ADSL modem system [11] using a top-down methodology.

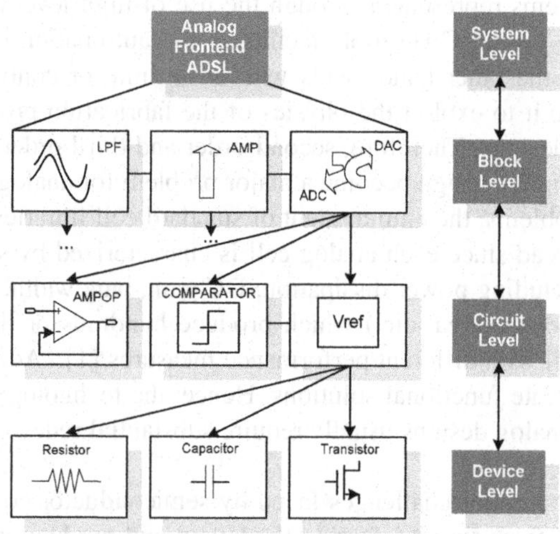

Fig. 1.3 Design of an analog frontend ADSL using a typical design flow

The main analog functions are high speed digital-to-analog (D/A) and analog-to-digital (A/D) conversion, low pass filtering (LPF) and transmitter and receiver gain-controlled amplifier (AMP). The design of the above system can follow two different approaches: design the circuit as a whole (not a recommended practice considering the systems dimension and complexity) or using a divide-to-conquer strategy (a common procedure to solve hard problems in different engineering domains). Applying this strategy, to analog integrated circuit design, results in a hierarchical subdivision of the system under study, in sub-blocks of different abstraction layers. Therefore, in the first level the initial system concept is refined in a series of steps, which will compose the building blocks of the next level: filters, data converters, amplifiers, etc, in the third level: amplifiers, comparators, etc. This decomposition process continues down until the device level. Moreover, the performance specifications obtained by these sub-blocks must be once more simulated to verify their correctness – this is known as the verification or validation phase. The validation phase can result in the redesign of the same block or a modification of the dependent block at the upper level.

One of the main advantages pointed out by the hierarchical methodology is the possibility of reusing the design knowledge [12][13][14]. Reusability of knowledge and procedures that are acquired by the design and refined for the verification of a building block in a hierarchical decomposition of a certain circuit can be used later when designing another circuit or used as a starting design for the next generation of the product. Moreover, there is an opportunity to perform system architectural exploration in order to improve the overall system optimization (e.g. finding an architecture that consumes less power) at a higher level before starting detailed circuit implementations [15]. In conclusion, embracing a top-down design representation is an important step to decrease the redesign spins and time-to-market, and to increase the levels of productivity.

1.2.3 Analog IC Design Flow

Besides the benefits of design methodology described in the preceding section, in the analog domain the hierarchy levels of this design process are not well de-fined and are not generally accepted [12]. Nevertheless, some approaches have already been implemented through this concept [16]. A general design flow for analog/mixed-signal systems is illustrated in Fig. 1.4 and is described through the following steps:

(a) System level – On this first stage of development, the required target specifications, technology process are defined. The overall architecture of the system is designed and partitioned into a set of high-level building blocks for the next level. During this phase, specifications for system are mapped into intermediate-level parameters which become the specifications for the lower level building blocks. The system-level partitioning and specifications are then verified using appropriate high-behavioral tools or system simulators such as Matlab [17], Verilog AMS [18], MMSIM [19], etc.

(b) Block level – In this stage there is an effective translation of the high-level building blocks into architecture of functional blocks required to realize the specified behavioral description. Then, all blocks are described individually in an appropriate hardware description language, like VHDL [20] and VHDL-AMS [21] and then verified against the specifications using behavioral simulations tools, such as Ultrasim [22], NcSim [22], Hsim [19], Modelsim [23], etc.

(c) Circuit level – For each analog building block an optimization process is provided, given the inherited specifications from the upper level and the selected technology process. The optimization is seen as an iterative process to determine physical dimensions at device-level. This stage covers two nuclear activities: the selection of the proper circuit topology and a device sizing methodology of the circuit parameters. A robust design should be achieved taking into account the process variations and device tolerances in order to guarantee a high yield design. The required performance specifications of the final circuit design are then verified using circuit simulations such as HSPICE [24] and Spectre [18].

(d) Layout hierarchy - In this stage the optimized building blocks obtained from the preceding step are mapped into a physical representation of the circuit schematic taking the form of a multilayer layout. Layout is a set of geometric shapes obeying design rules specified by the fabrication process. The layout area generated manually or automatically is optimized for minimum area. After the verification phase (verification of design rules (DRC)) layout is followed by the extraction of layout parasitics whose effects must be verified with circuit simulation in order to ensure that the initial performance does not deviate significantly from the target specifications even with their influence. Crosstalk, substrate coupling analysis and mismatch are also important subjects under the umbrella of layout techniques.

(e) Fabrication and Testing – In this last stage the masks are generated and the IC is finally produced. The fabrication process is accompanied by rigorous quality tests to avoid defective devices. The test and validation are fundamental steps to verify the correct operation of the circuit and so a good test board and test setup must be defined.

During the top-down path, each of these hierarchical abstraction levels is combined with a top-down and bottom-up strategy together with redesign or backtracking iteration loops [6],[12] as illustrated in Fig. 1.4 (right).

The top-down flow consists of the following steps:

(a) Topology selection – This step is responsible for choosing the best suitable circuit topology or architecture in order to meet the specification requirements from the preceding higher levels. There are several ways of solving this problem, from manual topology selection from a database employing heuristic rules [25] to the use of deterministic approaches, such as, the one that use the information from feasible performance space [26][27][28][29] or another one that combines the topology selection with the device sizing task using optimization based approach [30]. Section 2.1.1 presents a detailed discussion on methods applied to automate the topology selection task.

(b) Specification translation/Sizing – An optimized design is searched so that the complete block meets the required specifications. In higher levels of design hierarchy this process implements the decomposition of the block under design in a subset of specifications that are passed down in the hierarchy for each sub-block in such a way that the actual block meets its specs. For the lowest levels in the hierarchy, where the sub-blocks are materialized in single devices (transistors, resistors, etc), circuit sizing is taking place according to the performance specs and the selected topology received from upper levels. Here, there are two main approaches, namely, the knowledge-based approach and an optimization-based approach, relying on different optimization methods. In section 2.2 a detailed discussion on methods applied to automate this fundamental task is presented.

(c) Synthesis Verification – The optimized design is simulated and verified to see if performance meets the original requirements. If the desired performance is obtained, the design progresses down to sub-blocks of lower level. If not, a redesign process is initiated inside the same hierarchical level or a backtracking iteration is started involving other hierarchical level.

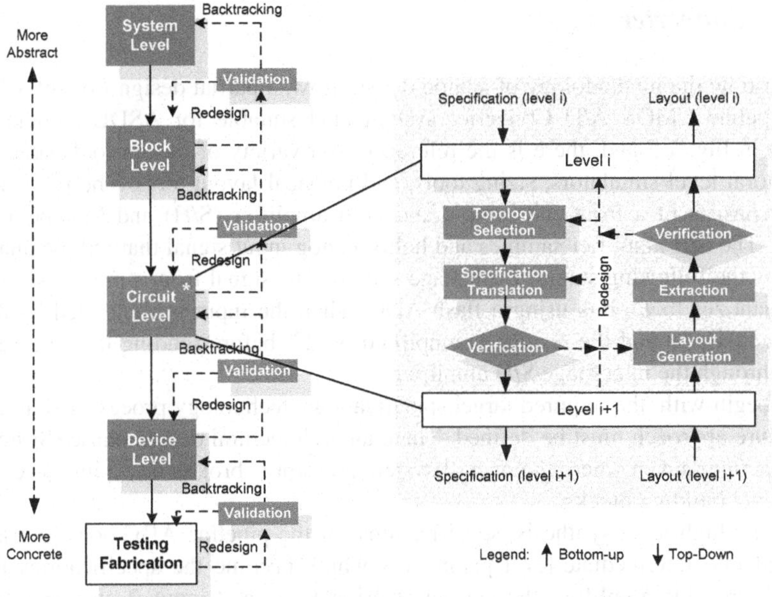

Fig. 1.4 The design flow in level i of analog IC design [6]

The bottom-up layout flow implements the next steps:

(a) Layout generation – It generates the optimal geometrical layout of the block under study taking into account design constraints.
(b) Extraction – After design rule checking (DRC) and layout versus schematic (LVS), layout is extracted to obtain the layout-induced effects to the circuit schematic, the layout parasitics.
(c) Layout Verification – The extracted layout is then verified and simulated to check the impact of the layout parasitics on the overall circuit performance. If the influence of parasitics produces unacceptable deviations from desired

performance, the design should be repeated again changing the dependent blocks in one or several upper levels. Some iteration among the upper levels in the design hierarchy is likely to occur. These verifications are done using 3rd generation SPICE simulators as HSIM [19] and NANOSIM [19] which can deal with large circuits in a reasonable period of time.

In order to support the process with adequate evaluation of design alternatives at different levels of abstraction, CAD (Computer Aided Design) tools have been developed, which reduce the execution time and allow the identification of problems at an early stage so that the right decision in the design process can be taken.

1.2.4 Analog Design Flow of a 15-Bit Pipeline CMOS A/D Converter

To illustrate the methodology of analog design flow, a partial design flow of a 15-Bit Pipeline CMOS A/D Converter system [11] suitable for ASDL modem is shown in Fig. 1.5 and, there is the reference to a variety of design tools such as behavioral level simulators, sizing tools, and physical layout tools. The pipelined ADC consists of a front-end sample and hold amplifier (S/H) and M pipelined stages. The S/H amplifier samples and holds analog input signal that will be quantized by the following stages. Each stage samples the signal from a previous stage and quantizes to *Bn*-bits using a flash ADC. Then the input is subtracted by the quantized signal and the residue is amplified by 2^{Bn} before sending it to the next stage through the inter-stage S/H amplifier.

To begin with, the required target specifications, technology process and an architecture approach must be defined. Then, an architectural design phase (System level) is undertaken where the overall system concept is broken down into a set of high-level building blocks.

During high-level synthesis, specifications for the pipeline ADC converter are mapped into intermediate-level parameters which become the specifications for the S/H, residue amplifier, the current steering DAC and input flash converter. The low-level synthesis phase uses these specifications as constraints to design the sub-blocks. These blocks are further decomposed (block level) until they are small enough in size to be treated as atomic circuit cells. Once all cells are designed at the circuit level, the system layout is done (layout level). Then, bottom-up verification is performed, and changes are made as necessary. During this process some iteration among the levels in the design hierarchy is likely to occur.

Special emphasis has been put on top-level simulations, to ensure the correct functionality of the entire converter circuit. Top-level simulations were performed using Matlab/Simulink [17] models, including all digital and analog blocks. Lower level sub-block simulations have been carried out using Spectre [18] and HSPICE [24] to verify circuit performance. Additional simulation with PowerMill [19] was taken to ensure the correct operation of the combination of digital and analog blocks.

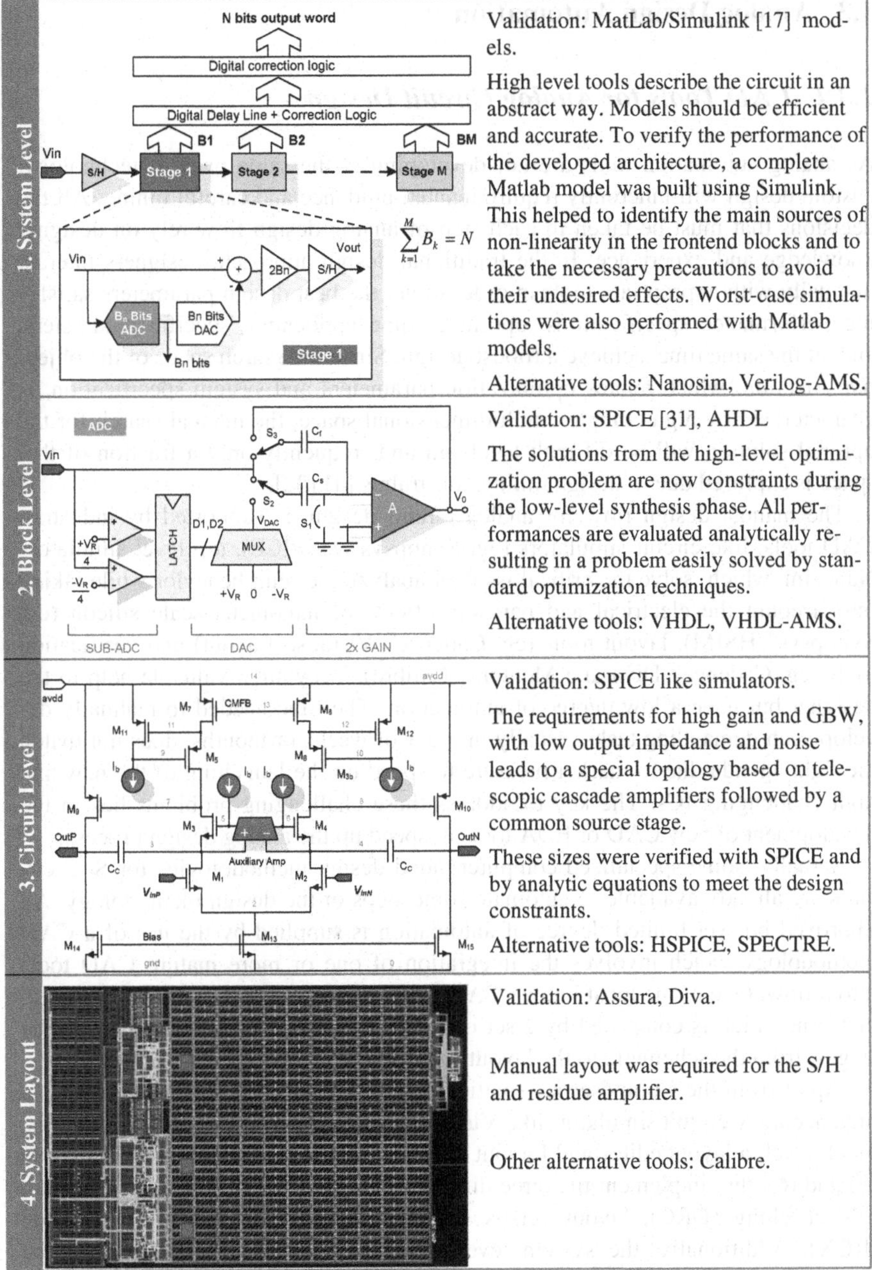

1. System Level
N bits output word
Digital correction logic
Digital Delay Line + Correction Logic
B1
B2
BM
Vin
S/H
Stage 1
Stage 2
Stage M
Vout
2Bn
Bn Bits ADC
Bn Bits DAC
Bn bits
Stage 1
$\sum_{k=1}^{M} B_k = N$
Validation: MatLab/Simulink [17] models.
High level tools describe the circuit in an abstract way. Models should be efficient and accurate. To verify the performance of the developed architecture, a complete Matlab model was built using Simulink. This helped to identify the main sources of non-linearity in the frontend blocks and to take the necessary precautions to avoid their undesired effects. Worst-case simulations were also performed with Matlab models.
Alternative tools: Nanosim, Verilog-AMS.
2. Block Level
ADC
LATCH
MUX
D1,D2
SUB-ADC
DAC
2x GAIN
Validation: SPICE [31], AHDL
The solutions from the high-level optimization problem are now constraints during the low-level synthesis phase. All performances are evaluated analytically resulting in a problem easily solved by standard optimization techniques.
Alternative tools: VHDL, VHDL-AMS.
3. Circuit Level
avdd
CMFB
Auxiliary Amp
OutP
OutN
Bias
gnd
Validation: SPICE like simulators.
The requirements for high gain and GBW, with low output impedance and noise leads to a special topology based on telescopic cascade amplifiers followed by a common source stage.
These sizes were verified with SPICE and by analytic equations to meet the design constraints.
Alternative tools: HSPICE, SPECTRE.
4. System Layout
Validation: Assura, Diva.
Manual layout was required for the S/H and residue amplifier.
Other alternative tools: Calibre.

Fig. 1.5 Hierarchical analog design flow of a 15-Bit Pipeline CMOS A/D converter

1.3 Analog Design Automation

1.3.1 CAD Tools for Analog Circuit Design

As analog automation tools are not developing at the same pace of technology, custom design will inherently require manual guidance and careful tuning. All the decisions that must be taken in each step of analog design flow rely on designer knowledge and experience. In the traditional design approach, designers interact manually with appropriate tools in order to get the best design parameters satisfying performance specifications, optimize some application specific parameters, and, at the same time, achieve a robust design. Since the search space of the objective function, which relates optimization parameters and system specification, is characterized by high complex multidimensional space, the manual search for the optimal solution will be difficult to obtain and frequently only a fraction of that space is explored due to design timing constraints [4],[32].

The manual design flow for analog circuit design, is supported by industrial CAD tools, like circuit simulators (eg. Synopsys® HSPICE), top level simulators Nanosim, which solve the critical issue of analyzing circuit behavior while taking into account the electrical and parasitic effects of nanometer-scale silicon (eg. Synopsys® HSIM), layout tools (eg. Cadence® Virtuoso Layout) and verification tools (eg. Cadence® Diva and Mentors® Calibre). They are a valuable help to the designer but have a low degree of automation. The time needed to manually develop such demanding tasks, usually in order of weeks or months, does not match the tight agenda due to market pressure to speed up the launching of the new and high challenging ICs. The key to address these challenging problems lies in the development of new CAD or EDA tools to speed up the analog design process.

Actually, some specialized computer-aided design methodologies for SoC circuits are already available to automate some steps of the design methodology. An improved but yet limited degree of automation is supplied by the use of a CAD methodology which involves the integration of one or more mature CAD tools into a flow. One of the most known CAD methodologies is the Cadence® Virtuoso platform which is composed by a set of integrated circuit tools that cover all the stages, from the schematic to the layout (see Fig. 1.6).

Apart from the Composer schematic editor (2), Cadence® Virtuoso includes a high accuracy circuit simulator, like Virtuoso Spectre (4) that is usually used at the block level, a layout editor and Layout Verification tools such as Assura and Diva (5) and (6), that implement the three different phases of layout process: the design rule checking (DRC), layout versus schematic (LVS) and parasitic extraction (RCX). Additionally, the system level analog behavioral descriptions may be simulated with the Verilog®-A simulator (AHDL).

The use of design management platforms, with a set of integrated CAD tools and database facilities to deal with the design transformations from the conception to the physical implementation, can significantly speed up the design process and

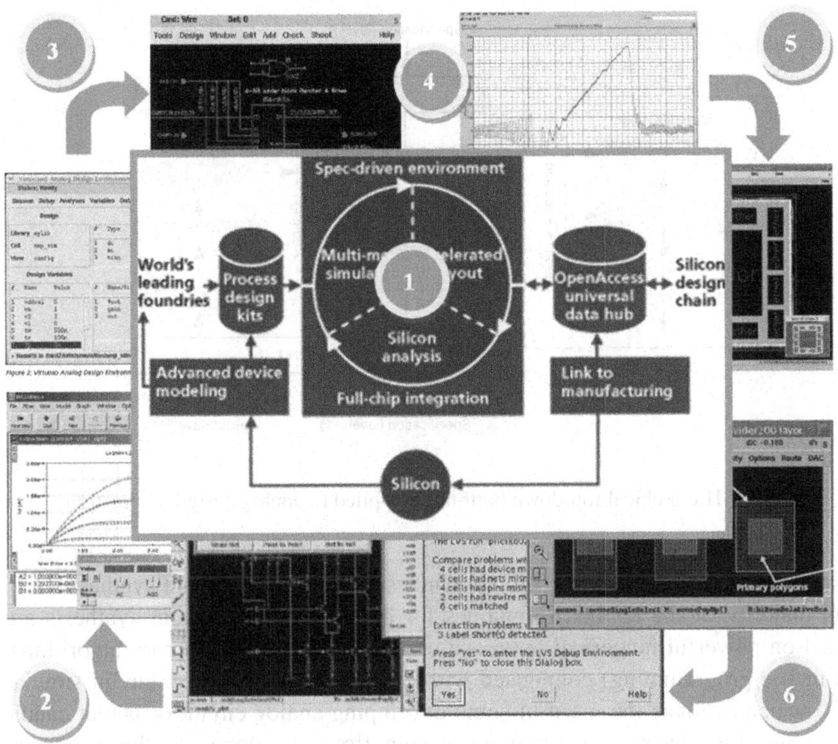

Fig. 1.6 CADENCE - Virtuoso custom design platform diagram

enhance the productivity of analog/mixed-signal design teams. These design management platforms are a valuable help in analog integrated circuit design but they are still far behind the development stage of design automation tools already available for digital design.

A new class of CAD tools for analog IC design have emerged taking advantage of the automation of some analog design process tasks. Design Automation (DA) tools help the automation of particular design tasks (Fig. 1.7), like a decision-making algorithm for circuit sizing (a system, a module or a cell), topology selection and layout generation, as well as the automation of retargeting operations and design flow control. The aim is to free the designer to more creative design tasks (working out better architectures and topologies) and to develop more efficient and accurate design automation tools.

1.3.2 Automated Analog IC Design

Today, the development of new analog synthesis tools is accomplished in order to fulfill the needs of the modern analog IC design. Analog synthesis consists

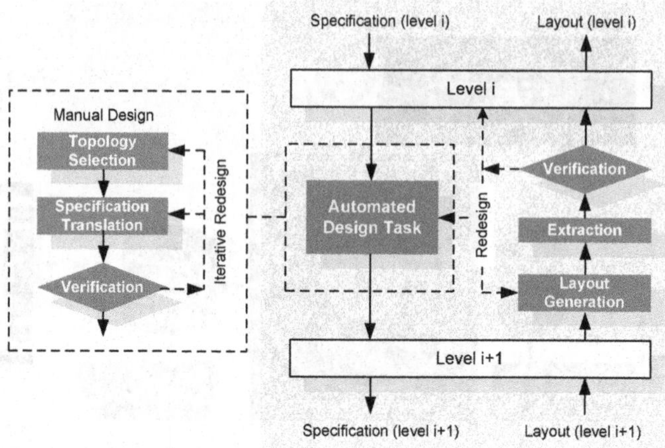

Fig. 1.7 Hierarchical top-down bottom-up applied to analog design automation

essentially of two major steps, circuit synthesis followed by layout synthesis [15]. The majority of the applied techniques for both circuit and layout synthesis are based on powerful numerical optimization engines (e.g. evolutionary algorithms, geometric programming) conjugated with evaluation engines (e.g. circuit simulators) which evaluate the merit of some developing analog circuit or layout candidate. Fig. 1.8 illustrates a top-down design flow, developed in this thesis, to produce a sized circuit based on certain circuit specifications.

The lack of an unique and structured design flow definition and, on the other hand, the mature state reached by the simulation/analysis support tools, led to the appearance of several simulation environments, including important commercial solutions, which showed promising results. However, the most relevant ones are still at very low level of abstraction. These tools focus predominantly the topology selection, circuit sizing and layout generation tasks since they are the most time-consuming processes. Examples of these automated EDA tools developed mainly by the scientific community domain are IDAC [33], ANACONDA [34], MAELSTROM [35], FRIDGE [36], AMGIE[37], GPCAD [38], [39], TAGUS [40], [41] and some layout generators like ILAC [42], ANAGRAM [43], [44], LAYLA [45], ALADIN [46] and LAYGEN [47], [48], among others. However, only very few examples in analog IC design industry embrace these new paradigms like Barcelona Design® and NeoLinear (Cadence®), which support both an automated circuit synthesis and layout generation tools called NeoCircuits [49] and NeoCell [49], respectively. Nevertheless, many of the used techniques were not efficient enough or produced the desired performance accuracy to become a mainstream application. In spite of these limitations, automated EDA tools can be helpful in increasing design productivity and circuit performances as shown symbolically in Fig. 1.9.

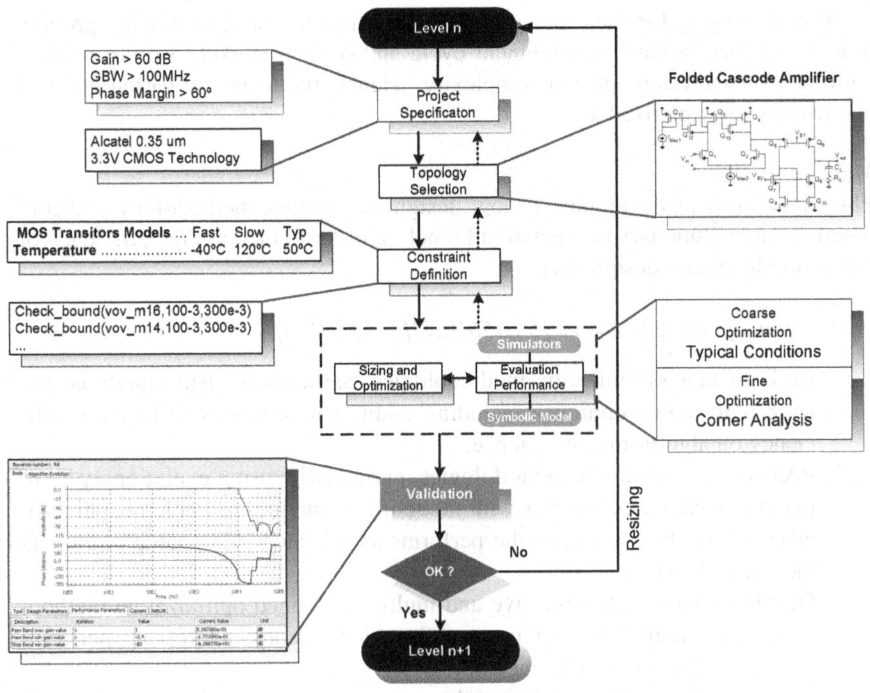

Fig. 1.8 Top-down design flow developed in this thesis

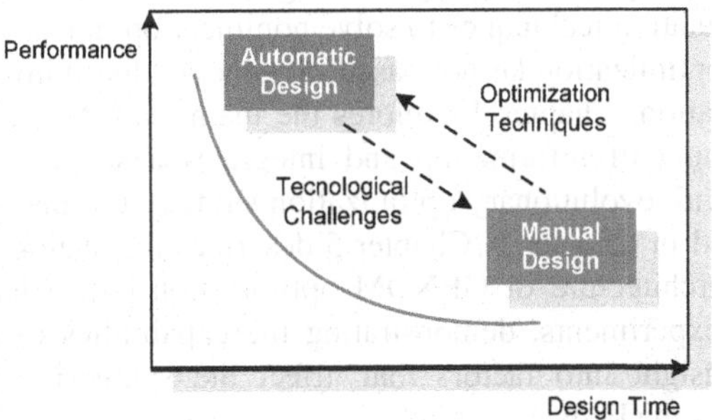

Fig. 1.9 Trends or trade-off in automatic design

In short, there is still a long way to go concerning IC design automation research, until the generalization of automatic analog IC design in industry takes place. The majority of the available tools do not provide a satisfactory answer to the complete design process, since they apply only to a specific part of the design this process. Therefore, there are still many situations, in which high performance analog IC design must be controlled by manual design rules, hence the research in design automation continues to be a matter of great importance.

From the exposed in the previous paragraphs there is a need to increase productivity, to reduce product development cycle and time-to-market in order to cope with the ever increasing design complexity. Hence, the main and partial goals of this work are summarized as:

Main goals [50]
This work aims at developing a new design automation methodology and tool based on an evolutionary computation kernel, in order to increase the efficiency of the analog IC circuit design cycle.

Partial goals [51][52][53][54][55][56][57][58][59][60]

1. R&D of new optimization methodologies, particularly, using modified genetic algorithm techniques including additional strategies to increase efficiency on algorithm convergence.
2. R&D of new synthesis methodologies, particularly, using evolutionary computation techniques together with modeling techniques to implement an efficient approach to optimize the performance of integrated analog circuits at the circuit level.
3. Develop a new multi-objective and multi-constrained optimization methodology for circuit sizing of integrated analog circuits, addressing performance, robustness and efficiency key factors.
4. Demonstrate the effectiveness of the new optimization methodology for well known analog integrated circuits and systems.

The remainder of the book is organized as follows: Chapter 2 presents an extensive state of the art analysis on Analog IC design Automation. Chapter 3 gives an overview on computation techniques to solve nonlinear optimization problems and introduces a new optimization kernel based on genetic algorithms applied to analog circuit optimization. Chapter 4 explores the main learning techniques used to manage large amount of information and integrates a support vector machine (SVM) approach with evolutionary optimization strategy for analog circuit design applications defined in Chapter 3. Chapter 5 describes the analog IC design environment and the architecture of GENOM optimization tool. Chapter 6 presents several synthesis experiments, demonstrating the capabilities of the system and providing some insight into factors that affect the synthesis process. Finally, Chapter 7 concludes the book.

References

[1] Leenaerts, D., Gielen, G., Rutenbar, R.A.: CAD solutions and outstanding challenges for mixed-signal and RF IC design. In: Proc. IEEE/ACM International Conference on Computer Aided Design, pp. 270–277 (2001)
[2] Gielen, G.: Modeling and analysis techniques for system-level architectural design of telecom front-ends. IEEE Trans. Microwave Theory and Techniques 50, 360–368 (2002)

[3] Lev, L., Razdan, R., Tice, C.: It's about time – Charting a course for unified verification. EETimes eeDesign News (2000), `http://www.eetimes.com` (Accessed March 2009)
[4] Horta, N.C.: Analog and mixed-Signal IC design automation: Synthesis and optimization overview. In: Proc. 5th Conference on Telecommunications, Tomar, Portugal (2005)
[5] Hasting, A.: The Art of Analog Layout. Prentice-Hall, Englewood Cliffs (2001)
[6] Gielen, G., Rutenbar, R.A.: Computer-aided design of analog and mixed-signal integrated circuits. IEEE Proceedings 88(12), 1825–1854 (2000)
[7] Sommer, R., Malcovati, P., Maloberti, F., Schwarz, P., Noessing, G., et al.: From system specification to layout: Seamless top-down design methods for analog and mixed-signal applications. In: Proc. Design Automation and Test in Europe Conference and Exhibition, pp. 884–891 (2002)
[8] Chang, H., Sangiovanni-Vincentelli, A., et al.: A top-down, constraint-driven design methodology for analog integrated circuits. In: Proc. IEEE Custom Integrated Circuits Conference, pp. 841–846 (1992)
[9] Toumazou, C., Makris, C.: Analog IC design automation: Part I - Automated circuit generation: New concepts and methods. IEEE Trans. Computer-Aided Design 14, 218–238 (1995)
[10] Donnay, S., et al.: Using top–down CAD tools for mixed analog/digital Asics: A practical design case. Kluwer Int. J. Analog Integrated Circuits Signal Processing 10, 101–117 (1996)
[11] Guilherme, J.: Architectures for high dynamic range CMOS pipelining analog to digital signal conversion. PhD dissertation, Dept. Electrical and Computer Engineering, Instituto Superior Técnico, Lisboa, Portugal (2003)
[12] Castro-Lopez, R., Fernandez, F.V., Guerra-Vinuesa, O., Vazquez, A.: Reuse Based Methodologies and Tools in the Design of Analog and Mixed-Signal Integrated Circuits. Springer, Heidelberg (2003)
[13] Liu, D.: A Framework for Designing Reusable Analog Circuits. PhD dissertation. Stanford University, Stanford (2003)
[14] Dastidar, T.R., Chakrabarti, P.P., Ray, P.: A synthesis system for analog circuits based on evolutionary search and topological reuse. IEEE Trans. Evolutionary Computation 9(2), 211–224 (2005)
[15] Gielen, G.: CAD tools for embedded analogue circuits in mixed signal integrated systems on chip. In: IEE Proc. Computers and Digital Technique, vol. 152(3), pp. 317–332 (2005)
[16] Chang, H., Malavasi, E., Sangiovanni-Vincentelli, A., Gray, P.R., et al.: A top-down, constraint driven design based generation of nbit interpolative current source D/A converters. In: Proc. IEEE Custom Integrated Circuits Conference, pp. 369–372 (1994)
[17] MATLAB, The language of technical computing. The MathWorks Inc. (1996)
[18] Cadence Inc, SPECTRE simulator and other cadence products (2009), `http://www.cadence.com/products` (Accessed March 2009)
[19] Synopsys Inc, Products and solutions-HSIM, PowerMill, NanoSim (2009), `http://www.synopsys.com` (Accessed March 2009)
[20] VHDL, IEEE standard VHDL language reference manual. IEEE Std 1076-2000 (2000)
[21] VHDL-ALS: IEEE standard VHDL analog and mixed-signal extensions reference manual. IEEE Std 1076.1 (2000)

[22] Cadence Inc, Products: Composer, Virtuoso, DIVA, NeoCircuit, NeoCell, UltraSim, NcSim (2009), http://www.cadence.com (Accessed March 2009)
[23] Mentor Graphics Corp, Products: Calibre (2009), http://www.mentor.com/products (Accessed March 2009)
[24] Synopsys Inc, HSPICE simulator (2009), http://www.synopsys.com/products/mixedsignal (Accessed March 2009)
[25] El-Turky, F., Perry, E.: BLADES: An artificial intelligence approach to analog circuit design. IEEE Trans. Computer Aided Design 8, 680–692 (1989)
[26] Harjani, R., Shao, J.: Feasibility and performance region modeling of analog and digital circuits. Kluwer Int. J. Analog Integrated Circuits Signal Processing 10, 23–43 (1996)
[27] Stehr, G., Graeb, H., Antreich, K.: Performance trade-off analysis of analog circuits by normal-boundary intersection. In: Proc. Design Automation Conference, pp. 958–963 (2003)
[28] Smedt, B., Gielen, G.: WATSON: Design space boundary exploration and model generation for analog and RFIC design. IEEE Trans. Computer-Aided Design of Integrated Circuits and Systems 22(2), 213–224 (2003)
[29] Smedt, B., Gielen, G.: HOLMES: Capturing the yield-optimized design space boundaries of analog and RF integrated circuits. In: Proc. of Design Automation and Test in Europe Conference and Exhibition, pp. 256–261 (2003)
[30] Kruiskamp, W., Leenaerts, D.: DARWIN: CMOS opamp synthesis by means of a genetic algorithm. In: Proc. ACM/IEEE Design Automation Conference, pp. 550–553 (1995)
[31] SPICE3 Berkerley, SPICE3 reference manual. University of Cincinnati (1993)
[32] Hjalmarson, E.: Studies on design automation of analog circuits – the design flow. PhD dissertation, Institute of Technology, Linköpings University (2003)
[33] Degrauwe, M., et al.: IDAC: An interactive design tool for analog CMOS circuits. IEEE J. Solid-State Circuits 22, 1106–1115 (1987)
[34] Phelps, R., Krasnicki, M., Rutenbar, R.A., Carley, L.R., Hellums, J.: ANACONDA: Simulation-based synthesis of analog circuits via stochastic pattern search. IEEE Trans. Computer-Aided Design of Integrated Circuits and Systems 19(6), 703–717 (2000)
[35] Krasnicki, M., Phelps, R., Rutenbar, R.A., Carley, L.R.: MAELSTROM: Efficient simulation-based synthesis for custom analog cells. In: Proc. ACM/IEEE Design Automation Conference, pp. 945–950 (1999)
[36] Medeiro, F., et al.: A Statistical optimization-based approach for automated sizing of analog cells. In: Proc. ACM/IEEE Int. Conf. Computer Aided Design, pp. 594–597 (1994)
[37] Gielen, G., et al.: An analog module generator for mixed analog/digital ASIC design. Wiley Int. J. Circuit Theory Applications 23, 269–283 (1995)
[38] Hershenson, M., Boyd, S., Lee, T.: GPCAD: A tool for CMOS op-amp synthesis. In: Proc. IEEE/ACM Int. Conf. Computer-Aided Design, pp. 296–303 (1998)
[39] Hershenson, M.M., Boyd, S.P., Lee, T.H.: Optimal design of a CMOS Op-Amp via geometric programming. IEEE Trans. Computer-Aided Design 20(1), 1–21 (2001)
[40] Horta, N.C., Franca, J.E.: High-Level data conversion synthesis by symbolic methods. In: Proc. IEEE Int. Symposium on Circuits and Systems, vol. 4, pp. 802–805 (1996)

[41] Horta, N.C.: Analogue and mixed-signal systems topologies exploration using symbolic methods. In: Proc. Analog Integrated Circuits and Signal Processing, vol. 31(2), pp. 161–176 (2002)
[42] Rijmenants, I., Schwarz, Y.R., Litsios, J.B., Zinszner, R.: ILAC: An automated layout tool for CMOS circuits. IEEE Journal of Solid-State Circuits 24(2), 417–425 (1989)
[43] Cohn, J., Garrod, D., Rutenbar, R.A., Carley, L.R.: KOAN/ANAGRAM II: New tools for device-level analog placement and routing. IEEE J. Solid-State Circuits 26, 330–342 (1991)
[44] Carley, L., Georges, G., Rutenbar, R.A., Sansen, W.: Synthesis tools for mixed-signal ICs: Progress on frontend and backend strategies. In: Proc. Design Automation Conference, vol. 33, pp. 298–303 (1996)
[45] Cory, W.: Layla: A VLSI Layout Language. In: Proc. 22nd ACM/IEEE Conference on Design Automation, pp. 245–251 (1985)
[46] Zhang, L., Kleine, U.: A novel analog layout synthesis tool. In: Proc. IEEE Int. Symposium on Circuits and Systems, vol. 5, pp. 101–104 (2004)
[47] Lourenço, N., Horta, N.C.: LAYGEN – An evolutionary approach to automatic analog IC layout generation. In: Proc. IEEE Conf. on Electronics, Circuits and System, Tunisia (2005)
[48] Lourenço, N., Vianello, M., Guilherme, J., Horta, N.C.: LAYGEN – Automatic layout generation of analog ICs from hierarchical template descriptions. In: Proc. IEEE Ph. D. Research in Microelectronics and Electronics, pp. 213–216 (2006)
[49] Cadence Inc. Products: Composer, Virtuoso, DIVA, NeoCircuit, NeoCell, UltraSim, NcSim (2009), `http://www.cadence.com` (Accessed March 2009)
[50] Barros, M., Guilherme, J., Horta, N.C.: Analog circuits optimization based on evolutionary computation techniques. Integration, the VLSI Journal, 136–155 (2010)
[51] Barros, M., Guilherme, J., Horta, N.C.: Analog circuits and systems optimization based on evolutionary computation techniques. In: Proc. Xth Int. Workshop Symbolic & Numerical Methods, Modeling and Application to Circuit Design, pp. 68–73 (2008)
[52] Barros, M., Guilherme, J., Horta, N.C.: An evolutionary optimization kernel using a dynamic GA-SVM model applied to analog IC design. In: Proc. 18th European Conference on Circuit Theory and Design, vol. 1, pp. 33–35 (2007)
[53] Barros, M., Guilherme, J., Horta, N.C.: GA-SVM feasibility model and optimization kernel applied to analog IC design automation. In: Proc. 17th ACM Great Lakes Symposium on VLSI, pp. 469–472 (2007)
[54] Barros, M., Guilherme, J., Horta, N.C.: GA-SVM optimization kernel applied to analog IC design automation. In: Proc. 13th IEEE International Conf. on Electronics, Circuits and Systems, pp. 486–489 (2006)
[55] Barros, M., Neves, G., Horta, N.C.: AIDA: Analog IC design automation based on a fully configurable design hierarchy and flow. In: Proc. 13th IEEE International Conf. on Electronics, Circuits and Systems, pp. 490–493 (2006)
[56] Barros, M., Neves, G., Guilherme, J., Horta, N.C.: An evolutionary optimization approach applied to analog circuit design. Poster presented at the 5th Conference on Telecommunications, Tomar, Portugal (2005)
[57] Barros, M., Guilherme, J., Horta, N.C.: An evolutionary optimization kernel with adaptive parameters applied to analog circuit design. In: Proc. International Symposium on Signals, Circuits and Systems, vol. 2, pp. 545–548 (2005)

[58] Barros, M., Guilherme, J., Horta, N.C.: GENOM2: An enhanced evolutionary approach to automatic synthesis matching designers methodology. In: 3rd Ph.D. forum at the Design, Automation and Test in Europe Conference, Munich, Germany (2005)
[59] Barros, M., Neves, G., Guilherme, J., Horta, N.C.: A distributed enhanced genetic algorithm kernel applied to a circuit/level optimization E-Design environment. In: Proc. Design of Circuits and Integrated Systems, pp. 20–24 (2004)
[60] Barros, M., Neves, G., Guilherme, J., Horta, N.C.: Enhanced genetic algorithm kernel applied to a circuit-level optimization E-Design environment. In: Proc. 10th IEEE International Conference on Electronics, Circuits and Systems, pp. 1046–1049 (2003)

2 State-of-the-Art on Analog Design Automation

Abstract. This chapter presents the State-of-the-Art (SOA) in analog circuit design automation. First, the analog design flow is reviewed and the fundamental trends in design automation are discussed. Then, the existing approaches to circuit sizing are presented, outlining in each case their advantages and limitations. Next, a detailed discussion over the existing tools approaches is provided. Finally, conclusions concerning the specification and design of a new analog design automation methodology implementation will be drawn.

2.1 Trends in Design Automation Methodology

A typical design flow for *analog and mixed-signal* IC circuits (AMS) consists of a series of design steps repeated from the system level to the device-level, and bottom-up for layout generation and verification. The steps between any two of these hierarchical levels are: topology selection, circuit sizing, design verification and layout generation task, illustrated in Fig. 2.1.

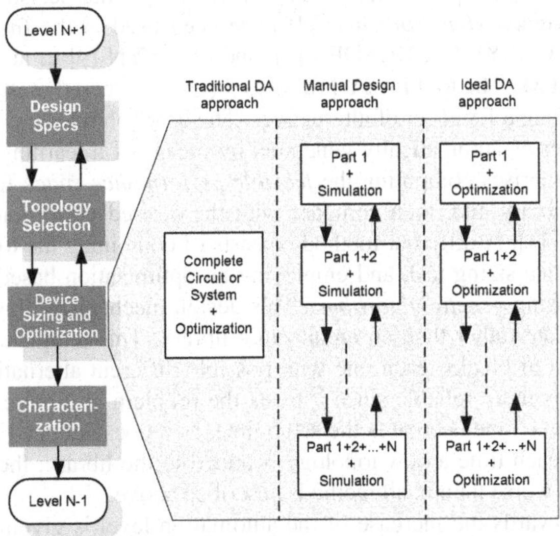

Fig. 2.1 Hierarchical level and design tasks of design flow architectures [1]

M.F.M. Barros et al.: Analog Circuits and Systems Optimization, SCI 294, pp. 19–47.
springerlink.com

In order to handle the increasing complexity of analog and mixed-signal IC design, a clear definition of a hierarchical design flow is essential. Despite the advances made during the last decades, the *design automation* (DA) tools in analog domain cannot support the complete design process, since they either concentrate on specific parts of the design flow or require the intervention of an expert designer. Moreover, they mainly address circuit level design as a whole (traditional design approach), which makes it difficult to apply to highly complex circuits and systems. Therefore, as the SoC complexity increases, the design automation tools must incorporate an hierarchical design decomposition feature in order to apply the well-known divide-to-conquer strategy already applied by most analog designers in a manual design approach.

Trends in this area have been running towards a class of design automation methodology under three aspects, improving:

- Flexibility, allowing the designer to have a higher interaction during the synthesis process and providing a more general approach to deal with multiple architectures or circuit types.
- Modularity, allowing the use of different tools and techniques to address different design tasks, such as topology selection, circuit sizing and layout.
- Hierarchy, allowing the handling of complex system designs and implementing strategies involving several abstraction levels.

2.1.1 Automated Topology Selection

The selection of an adequate architecture is fundamental to achieve a high performance design [2]. The topology selection task receives the performance specifications, for a particular class of circuits or systems, and delivers the most promising topology, traditionally from a predefined library. In IDAC [3] the decision is taken directly by the designer. *Heuristic rules* [4] have been used in the first attempts by TAGUS [5]-[6], OASYS [7], BLADES [8], and OPASYN [9] to automate the topology selection task. The tool FASY [10] uses *fuzzy–logic* based reasoning to select one topology among a fixed set of alternatives. The decision rules are introduced by an expert designer or automatically generated by means of a learning process. Another method comprises computing the *feasible performance space* for each topology within the library and, then compare with the desired performance specs, by AMGIE [2] and [11]. A different method consists of combining the topology selection with the device sizing task and employing an optimization based approach by DARWIN [12] using *genetic algorithms*. This design mechanism illustrated in Fig. 2.2, uses a template rather than an architecture library. This template specifies the topology in terms of blocks, each one with possible different alternatives. In short, this last method is more reliable since it treats the problem in a more deterministic way and at the same time decreases the setup time, as it does not need to rearrange a new set of rules each time a new topology is added to the library; the computation time, however, is worse than in all methods described above.

A new step towards the increase of the automation level is given by a new set of tools where topology selection is performed at a higher abstraction level. Instead of selecting the architecture from a library, a high level functionality of the system is defined now by a *hardware description language*. Then, an automatic translation is carried out, mapping the functional description into an internal representation and then into a specific topology. The mapping step is implemented after or during the device sizing process. This class of tools usually differs from the type of internal representation used. In the case of [13] the internal representation is a data flow graph, whereas in TAGUS [5]-[6], [14] and Konczykowska [15] it is a *symbolic signal flow graph* and in ARCHGEN [16] a *state-space description* is used. Then, a mapping operation is performed, resulting in a connection of lower-level building blocks whose parameters are optimized, obeying to some design constraints. The operation flow is executed in a top-down basis.

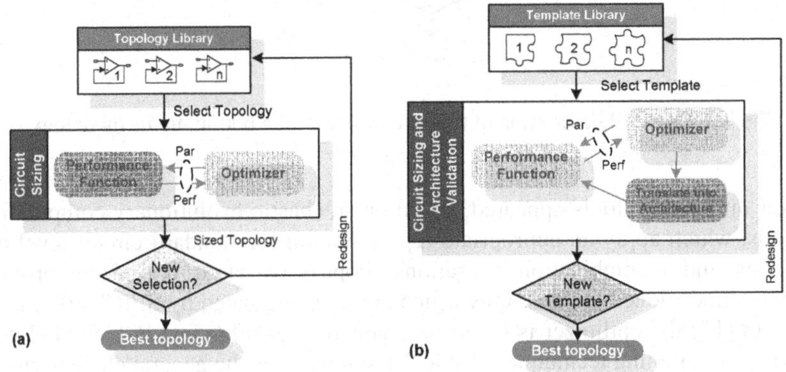

Fig. 2.2 Topology selection mechanism before (a) and during (b) device sizing

Finally, a design methodology able to create new topologies explores the immense potential from low abstraction level. Small elementary blocks are connected bottom-up to each other to form a new topology. The general description of this design methodology illustrated in Fig. 2.3 begins by selecting an initial topology, having in mind the desired specifications. As the design process takes place, an optimizer selects a transformation, adding or deleting a basic entity and/or attributing a value to a parameter. Various fundamental entities can be applied, such as, single transistors, elementary building blocks or node connections. As soon as the architecture is generated, the performance function is evaluated, providing some hints to the optimizer who makes a new selection of transformation. Essentially two exploration methods can be applied in topology generation for analog design. The knowledge-based exploration is based on a systematic or a random strategy where the circuit elements can be added, replaced or removed by an experienced designer with the help of standard CAD tools, like SPICE, and a circuit schematic editor. This method mimics the daily basis design approach supported mainly by simulation tools, and, therefore, suffers from the same drawbacks, i.e., as the number of entities in the system rise, the computational time increases accordingly. The computation time at the circuit description level can become intolerable if no efficient guidance is provided during the exploration step.

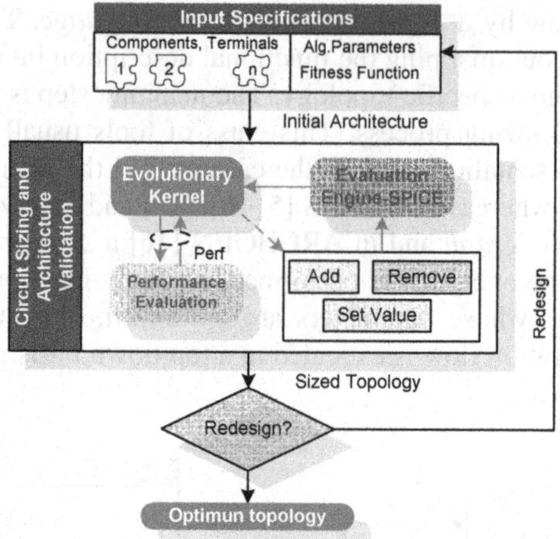

Fig. 2.3 A general description of the topology selection bottom-up methodology

New automation tools appeared, based on stochastic evolutionary computation methods, which apply an appropriate representation for standard circuit-level descriptions and recombination operations. Population-based optimizers provide multiple dimensioned architectures which are then simulated by SPICE-like simulators. In [17] the optimizer is based on a genetic algorithm and in [18]-[19] uses genetic programming techniques. Table 2.1 summarizes the general characteristics of automated topology selection and generation mechanisms.

Table 2.1 General characteristics of automated topology selection and generation

	Topology Selection		**Topology Generation**	
	Heuristic Rules	Feasible Region	Top-down*	Bottom-up
Tools	TAGUS [5]-[6], OASYS [7], BLADES [8], OPASYN [9] and FASY [10]	AMGIE [11], and Gielen [2], DARWIN [12]	Graeb [13], TAGUS[5]-[6], [14], Konczykowska [15]	Colombano[17], Koza[18], Toumazou [19]
Drawbacks	(-) Large set up time in order to update the selection rules to a new topology. (-) Qualitative approach and sometimes extremely difficult to codify heuristic rules.	(-) Time consumption	(-) Less generalized.	(-) Large time consumption. (-) No technological param. (-) No corner validation. (-) Not in a mature state.
Advant.	(+) Reduced execution time.	(+) Quantitative and general approach.	(+) Reduced execution time and well defined process	(+) Extremely promising. (+) Generic Approach. (+) No expert knowledge.

* Properties depend on methodology. This column considers knowledge-based approach.

2.1.2 Automated Circuit Sizing/Optimization

The sizing stage receives a topology description, a set of performance specs and a technology reference and, based on these inputs, produces a sizing solution for each block or component depending on the abstraction level. Several solutions were proposed derived from either knowledge-based methods, using some kind of knowledge and heuristics, or optimization-based approaches for both topology selection and specification translation or circuit sizing [1],[20]. The knowledge-based approach requires the expert knowledge of a designer to produce a set of rules and equations for every new circuit topology or technology. Another alternative is obtained considering the circuit sizing as an optimization problem. In these approaches the design problem is first mapped or modeled into an optimization problem and then solved by an appropriate optimization method, as illustrated in Fig. 2.4.

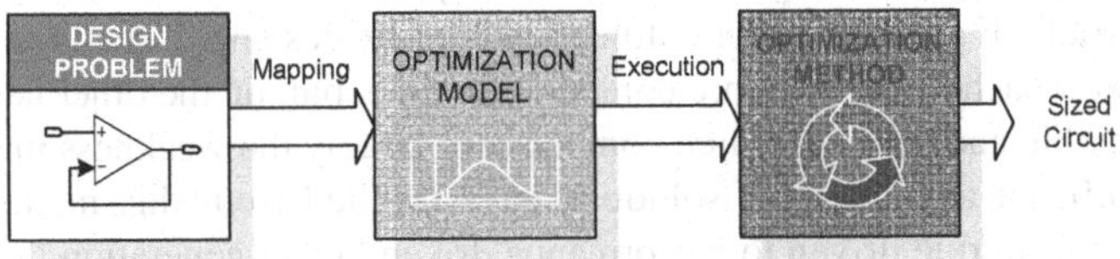

Fig. 2.4 Steps in optimization of circuit design

In this approach, there is a strong correlation between the modeling of a design problem and the way the modeled problem is solved. Since these steps are not independent and have influence on each other, the optimization method will be decided by the chosen model of the problem. For example, if the design problem is formulated in a set of *posynomial* equations the optimization method candidate could be the *geometric programming* (GP) algorithm or other computation algorithm able to process the convex optimization problem defined by posynomial equations [21]. If the design problem is formulated by SPICE models, a simulated annealing or a stochastic pattern algorithm could be used instead. Section 2.2 will explore the main optimization methods and alternative models in the area of analog IC design problems.

2.1.3 Automated Layout Generation

The earliest approaches to automate the layout generation followed a *procedural module generation* [22]-[23] with the codification of the entire circuit layout and its generation during the run time for the parameters attained during the sizing task. The procedural generators define a parametric representation of the geometric layout developed by the designer, accomplished either through a procedural language or a graphical user interface. The disadvantages of this approach are the lack of flexibility and generality and the high cost of the generation task.

Next, a *template-based approach* was developed [24] allowing the employment of geometric templates, which define the relative position and interconnection of devices. The templates are used to incorporate the designer knowledge into the optimization task. In spite of the low level of reusability achieved by procedural generators, the efficiency of this approach can be improved when user-defined templates are designed to be independent of both technology and specifications [4]. This approach is also suited when modifications in circuit parameters end in small adjustments to the global circuit layout structure, like technology migrations.

Later on the *optimization-based approaches* emerged, using optimization techniques to determine and predefine the layout solution. Small database of procedural cell generators, ANAGRAM [25]-[26], LAYLA [27] and ALDAC [28] synthesize an optimized layout configuration, searching the solution space formed by each cell layout positioning. The ALG [29] approach allows the generation of "optimal layout" of a circuit either automatically or by designer directives. On one hand these approaches require more computation time, but, on the other hand, they are more flexible and general, which compensates largely the weakness mentioned above. Significant technological solutions have resulted from this method [30]-[33], ranging from rule driven to performance driven layout generation tasks [27], reaching a more mature state when compared to what happens in the design automation tasks concerning circuit sizing [1]. The most frequent used optimization techniques in analog IC layout generation tools are simulated-annealing (ILAC [34], KOAN [25]-[26] and LAYLA [27]) and genetic algorithms (LAYGEN [35]-[36]). Simulated-annealing based approaches attained better results but lately the evolutionary approach has become a common option in many situations, like the hybrid solution defined by the genetic approach to simulated-annealing GASA [37] or the combined GA and Tabu Search (TS) used in [38] to develop a polycell placement algorithm for analog LSI chips. As both KOAN and LAYLA employ very simple cells on the database, some highly efficient structures, such as stacked or interdigited transistors, cannot be generated. Recent approaches, however, are tending to hybrid solutions employing optimization on blocks derived from knowledge-based systems. In the case of ALADIN [37],[39], the database usually relies on a hierarchical model where a cell is built using already defined cells. The use of compound cells reduces the search space because the number of cells handled during placement is lower and consequently reduces the computation times. Another knowledge-based approach with optimization is given by IPRAIL [40] and LAYGEN [35]-[36] in which the information presented in the template is defined manually or automatically and used to guide the layout generator during the synthesis procedure. The constraints defined in the template reduce the solution space, and allow the designer a higher control of the layout generation unlike the general optimization approaches [35]. Table 2.2 resumes the general characteristics of layout tools.

Table 2.2 Overview of layout tools

Tool	Year	Description	Techniques	Obs.
KOAN/ANAGRAM [25],[26]	1991	Macro-cell Place and Route; uses pre-defined small module generators data-base; synthesizes an optimized layout configuration from a given Spice netlist with symmetry, matching and tech. specs.	Optimization based with Simu lated Annealing.	The chosen library constitutes a limit of this method since an enormous number of pre-designed layout blocks is required
Layla [27]	1995	It takes into account symmetry constraints, performance degradation due to interconnect parasitics and device mismatches and combines this with geometrical optimization techniques (devices merges, abutment, etc.)	Optimization based with Simulated Annealing.	A performance-driven methodology where all performance constraints are satisfied. Optimize the layout quantifying the performance degradation.
A SKILLTM-based Library for Retargetable Embedded Analog Cores [32]-[33]	2001	Automatic generation and reusability of physical layouts of analog and nixed-signal blocks based on high-functionality pCells that are fully independent of technologies.	Knowledge-based	Parameterized cells (pCells) are organized hierarchically.
ALDAC [28]	2002	This tool providing means to generate multiple versions of full-stacked layout modules for the same circuit. The differences come from different MOS transistor splitting and grouping into stacks that can be performed either fully-automatically or user-controlled	Simulated Annealing	This approach minimizes parasitic diffusion capacitances of the circuit and permits economical post-layout simulation of multiple layouts for performance-driven
IPRAIL [40]	2004	Retargeting is achieved using an automatically extracted template and using a circuit optimizer to size the cells. It uses either a rule or a performance driven approach. It uses optimization based with knowledge-based.	Linear Programming and graph short path on the relational template extracted from the source layout	(+) General approach. (-) Larger run-time required.
ALADIN [25],[32]	2004	The layout generation is based on relatively complex sub-circuits. Designers can construct layouts of parameterizable modules in a technological and application independent way. The placement and routing of modules are performed automatically under the constraints defined by designer	Three phase Place e Route: 1 – GASA e half-perimeter routing 2 – VFSRA e global routing (fine tuning) 3 – Detailed Routing	Design platform for analog circuits, based on a user managed device generators library.

Table 2.2 (*continued*)

Tool	Year	Description	Techniques	Obs.
LAYGEN [35],[36]	2007	Expert knowledge is used to guide an evolutionary algorithm during the automatic generation of the layout The designer provides a high level layout description where position and interconnections are predefined. This template contains placement and routing constrains and is independent from technology. It deals with hierarchically templates for more complex circuits.	Knowledge-based with Evolutionary Computation techniques. Uses a geometric template.	(+) Speeds up retargeting operations or technology migration (-) Works better when changes in circuit parameters result in small adjustments. for the target technology
ALG [29]	2007	ALG is composed by three functional blocks: module generator, placer and router offering performance oriented layout generation in some of these blocks.	Cost function is a weighted sum function parasitics level, aspect ratio and mismatch, etc.	The user may choose the level of automation between full automation and user control.

2.2 Automated Circuit Synthesis Approaches

The computer-aided design methodology for AMS circuits foresees in a short-run the use of design automation tools to accomplish several tasks of the design methodology [41]. This trend began in 80's when the first automation tools applied to different tasks of analog design appeared like LAYLA [42] , IDAC [3],[22], DELIGHT.SPICE [43], BLADES [8] and OASYS [7]. The following sections review some of the most significant approaches for analog IC design including the knowledge-based, optimization-based approaches as well as the first commercial tools.

2.2.1 Knowledge-Based Approach

The knowledge-based approach presented, for instance, in programs like BLADES [8], IDAC [3], OASYS [7] and MDAC/ALSC [44]-[45], was the first to appear and is characterized by including a complete design plan describing how the circuit components must be sized to reach the solution for the design problem, even though, there is no guarantee of finding the optimum solution [2]. For example, the IDAC tool [3] takes advantage of the designer experience to manually derive or rearrange design plans to carry out the circuit sizing. OASYS [7] was built over a library of design plans defined for each elementary building block allowing the hierarchical representation of topologies, defined as the interconnection of several elementary building blocks. This system also implements a back tracking mechanism in order to recover from a malfunction implementation. The Fig. 2.5 illustrates the general design flow of knowledge-based approach.

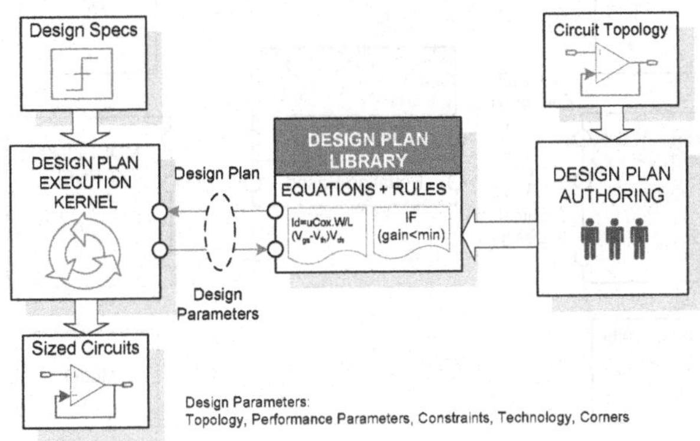

Fig. 2.5 Knowledge-based approach

In these methods, the main purpose is to encapsulate the designer's knowledge, building a pre-design plan with design equations and a design strategy that produce the component sizes in order to meet the performance requirements. This approach presents as major drawbacks the large overhead required to define a new design plan, the reformulation of the entire design plan when expanding the system to new topologies, and, finally, the migration to other technologies. Not only, it is a very time-consuming process to encode design knowledge for a given set of specifications, but design knowledge also has a limited lifetime. The rapid progress in process technologies made the acquired knowledge quickly out-of-date. Therefore, the application of these tools in industrial environments has been limited. However, after defining the design plan, the execution speed associated to the sizing procedure is extremely fast and the solution quality only depends on the models precision [1]. Naturally, this approach finds its applications restricted to small circuits or to more complex circuits but using simplified equations with the goal of achieving the first cut design.

2.2.2 Optimization-Based Approach

The optimization-based approach uses an optimization engine instead of a design plan to perform the design task. The optimization process is an iterative procedure where design variables are updated at each iteration until they achieve an equilibrium point.

The optimization-based approaches illustrated in Fig. 2.6, consist of an iterative loop, including an optimization engine or kernel together with an evaluation engine.

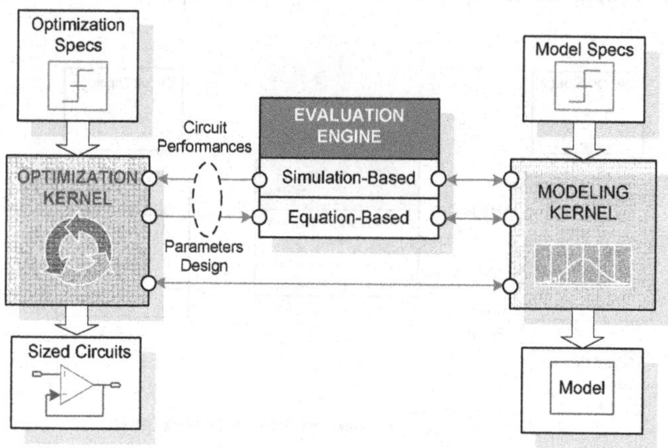

Fig. 2.6 Optimization-based approach

The optimization algorithm searches through the design space for values for each circuit component, whereas the performance evaluation tool verifies if the erformance constraints are met. If the system requirements are satisfied, then a solution is found and the component sizes are associated to the selected topology. The optimization engine should apply the appropriate techniques to efficiently guide the search mechanism in order to minimize the number of iterations required for the optimization process.

Different approaches can be described depending on the type of performance evaluation and the optimization technique employed. Concerning performance, the evaluation engine is typically implemented using an equation-based optimization, a simulation-based optimization or modeling-based optimization approach.

2.2.2.1 Equation-Based Methods

The equation-based methods use analytic design equations to evaluate the circuit performance. These equations can be derived manually or automatically by symbolic analysis tools. Then, the problem can be formulated as an optimization problem and normally solved using a numerical algorithm. Some of the most relevant approaches are OPASYN [9], STAIC [46], MAULIK [47], ASTRX/OBLX [48], AMGIE [11], GPCAD [49][50], SD-OPT [51]. This approach presents the advantage of allowing a performance evaluation speed-up (short evaluation time). The main drawback is that analytical models have to be used to derive the design equations for each new topology and, despite recent advances in symbolic circuit analysis [52]-[53], not all design characteristics can be easily captured by analytic equations. The approximations introduced in the analytic equations yields low accuracy designs especially in complex circuit's designs.

A promising methodology that has received much attention is related to circuit problems formulated in *posynomial form* (expression 2.1) and seen in tools like GPCAD and [21], [54]. This methodology solves the convex formulated problem by *geometric programming* techniques in a very short time. These techniques take advantage of the development of extremely powerful interior-point methods for general convex optimization problems [21],[50]. Besides the extreme efficiency of these methods they have another great advantage, as the global solution is always found, regardless of the starting point. However, a significant drawback still exists due to the difficulty to reformulate high-accuracy device models as posynomials equations, "performance specifications, and objectives that can be handled are far more restricted than any of the methods described above" [50]. Despite the progress presented in [54] the lack of an automated scheme to generate these equations limit the usage of this tool to a few, predefined, circuit structures.

$$f(x_1,\ldots,x_n)=\sum_{k=1}^{t} c_k x_1^{\alpha_{1k}} x_2^{\alpha_{2k}} \cdots x_n^{\alpha_{nk}} \tag{2.1}$$

$$where,\ c_j \ge 0 \text{ and } \alpha_{ij} \in \Re$$

2.2.2.2 Simulation-Based Methods

The simulation-based approaches such as DELIGHT.SPICE [43], FRIDGE [55], FASY [10], ANACONDA [56], MAELSTROM [57] and DARWIN [12] consist of using some form of simulation to evaluate the circuit's performance. In general, these types of tools for analog circuits design employ a circuit analysis tool in the inner loop of the optimization cycle to determine the circuit's performance. This is pointed out as a very flexible solution when compared with other methodologies (equation-based, knowledge-based) once it accommodates to any type of circuit topology and yields superior accuracy (depends on simulator models). Presently, the use of SPICE-like simulators are almost generalized and essential to support the optimization engine with all the feedback related to an accurate circuit evaluation, involving different performance characteristics, technological parameters and worst case corners analysis. Moreover, within this approach the same circuit can be optimized several times for different specs as long as the goal function is adapted, therefore, with this approach virtually all types of circuits can be sized and optimized with low setup time.

Despite these advantages, automated circuit sizing is not as commonly used as for example, circuit simulation, since it is computationally too expensive to evaluate electrical simulations. However, with the exponential increase of computer power and efficient use of optimization algorithms it has become increasingly favorable. Nevertheless, a key difficulty is that the analog design problem, with all the involved design knowledge and heuristics, has to be formulated as an optimization problem, which often presents a high threshold for using a circuit-sizing tool.

2.2.2.3 Learning-Based Methods

A step forward to enhance the efficiency of optimization based methods corresponds to the introduction of modeling techniques [58] based in *learning strategies*, which are clearly more time-efficient, during the optimization cycle. In this class of methods, the behavior of the circuit to be optimized is modeled by a learning mechanism based on the distribution of variation parameters, thus allowing a quick evaluation of the performance for a specific set of design parameters. Nevertheless, these methods require a set of training samples in order to build the model in the target region. Generally, a high accuracy evaluation engine is used, such as a circuit simulator to evaluate the performance of the training sample. The amount of the training data will influence the accuracy of the performance predictions made by the learning machine. However, an increase on the training data means that the evaluation of the performance will take more time. Like in equation-based methods, there will always be a trade-off between accuracy and efficiency.

Some of the most significant behavioral-based methodologies are described by Rutenbar [58], Alpaydin [59], Vincentelli [60] and Vemuri [61]. In the basis of Alpaydin tool is a neural-fuzzy model approach combined with an evolutionary optimization strategy and simulated annealing where some of the AC performance metrics are computed using an equation-based approach.

In [60] Sangiovanni-Vincentelli and al. use a learning tool based in support vectors machines (SVM) to represent the performance space of analog circuits. Based on the knowledge acquired from a training set, the performance space is modeled as mathematical relations translating the analog functionality. In this work two classes of SVM are confronted in an optimization-less strategy where additionally two improvements of the basic one-class SVM performances, conformal mapping and active learning, are proposed by enhancing the resolution in the support region boundaries. SVMs are trained with simulation data, and false positives are controlled based on a randomized testing procedure.

The Vemuri approach [61] presents a performance macro-model for use in the synthesis of analog circuits based in a neural network approach. On the basis of this mathematical model is a neural network model approach that, once constructed, may be used as substitute for full SPICE simulation, in order to obtain an efficient computation of performance parameter estimates. The training and validation data set is constructed with discrete points sampling over the design space. The work explores several sampling methodologies to adaptively improve model quality and applies a sizing rules methodology in order to reduce the design space and ensure the correct operation of analog circuits.

2.2.3 *Commercial Tools*

Besides the efforts introduced above some commercial EDA tools for circuit sizing have emerged in the past few years, such as the ADA's [63] Genius product line now integrated in Synopsis, Barcelona Design [49] which employ convex optimization techniques and recently the NeoCircuit from Neolinear Inc. [62], which implements a simulation-based approach.

The ADA (Analog Design Automation) Genius line of optimization tools, including Creative Genius, which automates device sizing and biasing to optimize circuit performance, and IP Explorer, which graphically provides N-dimensional circuit performance tradeoffs, were recently acquired [63] and integrated within the analog design environment from Synopsys [62], Mentor Graphics and other EDA vendors. The Genius tool builds its database of circuit from a transistor-level netlist, testbenches, objectives, process and environmental variations and variables. This system is comparable to NeoCircuit once it implements a simulation-based approach and interfaces with several industrial circuit simulators using parallel computation architecture.

The now extinct Barcelona Design was founded in 1999 by Stanford University researchers that apply advanced optimization techniques based in convex optimization to develop optimization solutions for a broad spectrum of circuit design problems including analog, RF and digital circuits. The final product introduces the synthesizable IP (intellectual property) block, which contains the required design equations written as posynomial expressions. The particularity of these products is that in opposition to standard IP blocks, which meet the given specifications, these blocks, may be synthesized to meet a range of different specifications. This implementation was reported to be able to increase design speed by 100 times and reduce total design costs by up to 50%.

The Neolinear package, now acquired by CADENCE [65], is composed by the NeoCircuit package, a simulation-based analog circuit sizing engine and the NeoCell module to automate the layout generation process. Both design packages together with the optimization engine based on a "genetic annealing" scheme creates a complete analog design flow. The integration of Neolinear's products in the Virtuoso design environment takes advantages of Cadence's multi-mode simulation and extensive layout design capabilities.

2.3 Design Automation Tools: Comparative Analysis

The existing design automation approaches are here compared, taking into account some qualitative and quantitative measures described in subsection 2.3.1. Table 2.3 presents the analog sizing tools used in this study and the conclusions are presented in the following subsections.

Table 2.3 Overview of analog sizing tools (1/4)

Tools \ Features	Date	Evaluation Class	Algorithm Techniques	Equation / Design Plan	Implemen. Language	Robust Design	Circuit Complexity	Computa-tion Time Effort	Setup Time	Interactive Design	Bookkeeping	Encapsula-tion	Advantages and Parti-cular Properties
IDAC [3]	1987	Knowledge Based (KB) Simplified Equations	Design plan + Post Opti-mization	Symbolic analyzer + manually	☒	☒	5-20 devices	A few seconds (VAX780)	Long time to add a new cir-cuit (months). Employs worst-case design.	☒	☒	☒	Able to size a library of analog schematics as a function of technology and building-block specs.
DELIGHT SPICE [43]	1988	Circuit Simulator with SPICE	Feasible Directions	☒	☒	✓	20 Par. 28 devices	18h (Masscomp MC 500)	Moderate setup	☒	☒	☒	Uses a large library of opt. algorithms. Problems with simulator convergence.
OASYS [7]	1989	KB Simplified Equations	Design Plan + Backtracking	Manually and specific to each class	☒	✓	19 Par. 17 devices OpAmp	3 sec. (VAXstation II/ GPX)	Long setup time (approx. 6 months), including circuit analysis	☒	☒	☒	Apply task decomposition, simplifies design reuse. Find out feasible and infeasible surfaces
BLADES [8]	1989	KB Lookup Tables and Rules	Artificial Intelligence. Uses divide and conquer strategy.	Manually	☒	☒	34 devices OpAmp	A few seconds to design and 20 min to complete simulation	Long time to create and maintain look-up tables.	☒	☒	☒	Uses a divide and conquer strategy to partition blocks written by "if-then clauses". Ability to perform circuit design, generate test files and invoke circuit simulator automatically

Table 2.3 Overview of analog sizing tools (*continued* 2/4)

Tools \ Features	Date	Evaluation Class	Algorithm Techniques	Equation / Design Plan	Implemen. Language	Robust Design	Circuit Complexity	Computa-tion Time Effort	Setup Time	Interactive Design	Bookkeeping	Encapsula-tion	Advantages and Parti-cular Properties
OPASYN [9]	1990	Simplified Equations	Grid + Steepest Descent	Manually. Sim-ple analytical model	C and Lisp	✓	7 Par., 60 devices OpAmp	5 min. (VAX 8800)	2 weeks for known circuits and a few days for simpler cir-cuits.	⧄	⧄	⧄	Interface with Berkeley CAD envi-ronment. Includes layout simul.
MAULIK [47]	1991	Simplified Equations + BSIM	Branch and Bound	Manually	⧄	⧄	39 Pararm., 19 devices	1 min. DEC 3100	6 months, in-cluding circuit analysis	⧄	⧄	⧄	Topology selection + device sizing
STAIC [46]	1992	Simplified Equations	Two Step Optimization	Manually	C++	⧄	22 device OpAmp	3 min. MIPS 2000	Entering new descriptions take long time.	⧄	⧄	⧄	Presents an analog description lan-guage with 4 levels
FRIDGE [55]	1994	Circuit Simulator	SA with SPICE +local Method	⧄	⧄	☒	10 Pararam OpAmp	45 min. Li-mited to a few thousand eval-uations.	Effort to add a circuit in about one hour.	⧄	⧄	⧄	Has only been dem-onstrated for prob-lems of a small number of opt. var.
SD-OPT [51]	1995	Equations + Behavioral Simulations	SA	Manually	⧄	☒	4th-order sigma-delta modulator	1.5 weeks	Exhaustive analysis Required. High cost to imple-ment new structure	☒	☒	☒	Design for switch-ed-capacitor delta-sigma modulators
FASY [10]	1995	Simulation Based	Fuzzy + SA + Gradient+ NN	SPICE as the evaluation en-gine.	⧄	☒	9 devices OpAmp	6 hours (96 MIPs worksta-tion)	⧄	☒	☒	⧄	Hybrid SA for coarse and gradient for fine simula-tions. Fuzzy logic for topology select.

Legend: ⧄=There was no mention of it ☒=Means "not implemented in tool" ✓= Means "implemented in tool"

Table 2.3 Overview of analog sizing tools (*continued* 3/4)

Tools \ Features	Date	Evaluation Class	Algorithm Techniques	Equation / Design Plan	Implemen. Language	Robust Design	Circuit Complexity	Computation Time Effort	Setup Time	Interactive Design	Bookkeeping	Encapsulation	Advantages and Particular Properties
ISAID [66]-[68]	1995	Simplified Equations, Qualitative Reasoning	Qualitative Reasoning + Post Optimization	Manually	⧄	⧄	8 Par., 13 devices OpAmp	⧄	⧄	⧄	⧄	⧄	Replace exact performance relations with quantitative relations.
ASTRX-OBLX [48]	1996	AWE + Equations	SA	Automated. Setup highly automated	C	⊠	33 devices	11.8h (IBM RS/6000-550)	A few days to add new circuits.	⊠	⊠	✓	Simulation + equation Based on waveform evaluation.
GPCAD [48],[55]	1998	Simplified Equations + Geometric Progr. (GP)	Simple Primal Barrier Method	Equations for Opamps are generated manually.	MAT-LAB	⊠	10 devices OpAmp	Very fast run time, in the order of a few seconds	Doesn't specify setup time. Doesn't include automatic generations of equations.	⊠	⊠	⊠	Achieves short execution times by the use of Geometric programming.
MEALSTROM [57]	1999	Circuit Simulator	GA+SA	⧄	C++	⊠	27 Par., 32 devices	3.6 h (15) SunSparc - 1)	Uses Cadence GUI. Long times are associated with simulations run.	✓	✓	✓	Uses simulator in distributed environment.
ANACONDA [56]	2000	Circuit Simulator	Stochastic Pattern Search	⧄	C++	⊠	20 Par., 37 devices	10 h (24 Sun Ultra 10)	Encapsulate commercial simulators (TISpice).	✓	✓	✓	Add constraints on devices to ensure a safety margin from manuf. or envir. variations. Parallelism.

Table 2.3 Overview of analog sizing tools (*continued* 4/4)

Tools \ Features	Date	Evaluation Class	Algorithm Techniques	Equation / Design Plan	Implemen. Language	Robust Design	Circuit Complexity	Computation Time Effort	Setup Time	Interactive Design	Bookkeeping	Encapsulation	Advantages and Particular Properties
AMGIE [11]	2001	Simplified Equations	SA + several Local Methods	Symbolic analyzer + manual	⧄	☒	14 Par 9 devices	5 min	8 hours	☒	☒	☒	Complete design flow. Device sizing is a compilation of tools
ALPAYDIN [59]	2003	Fuzzy Neural Network	Evolutionary Strategies + SA	Training points + circuit simulator	⧄	✓	31 devices OpAmp	45 min. (124 min. with mismatch model)	Model accuracy depends on training points. Cannot handle mismatch.	☒	☒	☒	The N-F model uses a set of training data from SPICE.
VINCENTELLI [60]	2003	SVM	Use LIBSVM Package.	Performance model of analog circuits	⧄	⧄	4 opt. variables	10's min 50.000 samples	⧄	⧄	⧄	⧄	Improve accuracy of estimator using an Active Learning strategy
VEMURI [61]	2004	NN	MIT GAlib	performance parameter macro-models	MAT-LAB	⧄	5-33 devices. Up 10 opt. var.	51.9 µs of execution time	It takes 1h47min in a SunBlade 100 to generate the model training samples(3125).	⧄	⧄	⧄	Explores sampling methodologies to improve model quality.
GENOM [69]-[70]	2006	Circuit Simulator +SVM	GA / SVM	SPICE/ HSPICE engines + "feasibility models"	C	✓	31 Par. 21 devices 41 const.	20 min. with robust design (Intel Core2 CPU @ 2.40GHz PC)	Encapsulate in-house environment (AIDA). Can use dynamic model generation.	✓	✓	✓	Apply learning strategies. Distributed processing and robust design.

Legend: ⧄=There was no mention of it ☒=Means "not implemented in tool" ✓= Means "implemented in tool"

2.3.1 *Specific Characteristics*

The tools described in Table 2.3 can be evaluated by several metrics that measure the final solution quality. The first column "date" is performance independent. There is no correlation between the availability of the design tool and its efficiency or accuracy. On the contrary, the next three characteristics columns "Evaluation Class", "Algorithm Techniques" and "Equation/Design Plan", which are often used for classification purposes, will have an important influence in the performance and accuracy as will be shown later in this chapter.

Particularly, the following metrics were considered in order to compare the characteristics of the presented applications.

(a) *Robust Design*: As far as sizing is concerned, robust design has to do with the accuracy and robustness of the solution. Accuracy is a measure of the quality that shows the difference between the synthesis tool's performance prediction mechanisms and the real performance of the obtained solutions, possibly including the layout-induced degradation. Robustness can be described as the capacity of the sizing tool to build and test circuits tolerant to manufacturing faults and operating point variations.

(b) *Automation Level*: It can be described as the ratio of time needed to accomplish the task of designing a circuit manually to the time spent on designing the same circuit with the help of a synthesis tool. In this metric two aspects must be considered:

- Run time response: The period of time taken by the optimization tool to give the first solution to the problem.
- Setup time: The setup time is a measure of the time spent by the designer to adequate the problem to the synthesis tool. This time is often longer than the execution of the synthesis tool. This feature is particularly important because it is strongly correlated with the success and acceptability of the tool. What is the advantage of a design tool which has the remarkable prodigy to output some results in seconds, if it is necessary two months to setup the complete algorithm of a hypothetic circuit when it is known it could be designed by hand in one month? Excluding a reused-based scenario, the answer is obvious "None".

(c) *Scope of the tool*: It can be described as a group of analog design problems, which can be solved by this tool. This is an important feature for analog design, because these problems usually require several types of optimization techniques. An analog synthesis tool which aims at solving a wide range of design problems will be successful in the long run, whereas tools planned to solve a narrow range of problems will soon be out of date. Although, it is not shown in Table 2.3, it will be used later for comparative analysis.

(d) *Design facilities*: It can be described as the set of additional features that can enrich a synthesis tool.

- **Multi-objective Optimization.** The DA tool presents the final solution in terms of a set of designs representing complementary tradeoffs of specific objectives (for example, area versus consumption) instead of single design response.

- **Interactive Design.** The tool optionally produces intermediate performance reports (in the form of text or graphics) throughout the design execution time to inform the IC designer on the optimization progress. At the same time, the IC designer optionally has the possibility to interact with the tool in real time manner to tune up some parameters, e.g., the dimension of a transistor or the redefinition of some design bias.
- **Bookkeeping Facilities.** The tool should have additional capacities to help with the introduction and management of all the necessary data including the management of different technological files, different classes of circuits (e.g., operational amplifiers, phase-locked-loops, etc.), different performance measurements, different design parameters, different components, different topologies, and so on.
- **Encapsulate Details.** Some tools interact with external programs and so it makes sense that the interface with these additional tools can be made in an automatically way hiding unused options.

2.3.2 Performance Analysis

Performance results are intrinsically correlated with several factors, like the evaluation engine, the search mechanism, the technological model precision, the computer platform used to run the application, etc.

The computation time is highly correlated with the nature of the evaluation engine. All approaches leading with models derived either by numeric equations or by some artificial learning machine method are able to reach solutions quickly, however, the quality of results are always estimated approaches and the solution quality only depends on the models precision [1]. This important trade-off between accuracy and computation time can be observed in Table 2.4. By contrast, simulation based methods that play with a high accurate circuit simulator in each optimization loop cycle are able to produce good quality results, but at the expense of higher execution times.

In the knowledge-based approaches the execution speed is the highest of all methods, considering that, the design plan is already defined. In equation-based approach, this value is normally high and is directly related to precision of the designed equations that need to be additionally introduced. The use of automatic methods to generate equations, like symbolic analyzers, can significantly reduce the input overhead and increase automation levels.

The *setup time* in equation and knowledge-based approaches is normally high and is directly related to the precision of the designed equations that need to be considered. The use of tools to generate equations, like symbolic analyzers, can significantly reduce the input overhead and increase automation levels. In the simulation based approach the level of interaction is the lowest of all methods. Only a few configuration parameters are necessary to setup the data for the external evaluation tool and the optimization algorithm. The behavioral-based approach is compared in performance with the equation based approach once both use fast evaluation models, but in what concerns to setup time, they behave like the simulation based one, requiring the configuration of only a small number of parameters. However, additional time will be required to configure the tool to produce the target samples. In this case the time to build the training points will augment the setup time slightly.

Table 2.4 Factors affecting tools performance

	Knowledge	Equation	Simulation	Behavioral
Computation Time	+	+	-	+
Setup Time	- -	- - M/-A	+	-
Accuracy	-	-	+	-
Robustness	-	-	+	-

M- means "by manual equation" and A-"automatic by symbolic methods"
Symbols ordered from the best to the worst: '+', '-', '- -'

With regard to robustness, the most promising classes of tools come from methodologies which are able to produce high accurate solutions like the simulation based approaches, although they require multiple simulations which adversely affect the run-time of the algorithm in a few orders of magnitude. Theoretically, all other approaches could reach the desired robustness in case they are able to produce efficiently accurate models. However, this solution would be impractical due to the large time spent in the preparatory phase to obtain those models. Besides that, the equation-based as well as behavioral-based approaches were explored in order to model the distribution of variation parameters in a form which can be efficiently optimized. However, the accuracy of these approaches is questionable.

2.3.3 Optimization Techniques

Analog circuit design is considered a hard optimization problem and has been used by researchers in classical artificial intelligence, classical optimization, and intelligent systems as a testbench for their methods. Some of the most significant approaches concerning the optimization-based techniques are presented in Table 2.5. However, both the classical AI approaches (tree search, expert systems, etc.) and the classical optimization approaches have some drawbacks. The former suffer from the lack of flexibility: a lot of effort is needed in order to handle new processes, topologies, etc., and even when those are in place, the tools tend to fail whenever slightly different problems are handled. The latter, tend to be gradient-based approaches, which can only be applied to local parameter optimization when the objective functions are differentiable and the design space is continuous.

Nevertheless, complex circuit problems tend to be non-differentiable and may have continuous or discrete design spaces making these approaches inefficient. The "Intelligent" systems-based approaches (EA+SA+Stochastic), on the other hand, offer the potential to meet the target required by an analog cell design in such complex search spaces. Through the observation of Table 2.5 the predominance of these methods for implementing the optimization engine is obvious.

Table 2.5 Optimization-based techniques

	EA	SA	Stochastic	SA+Local	AI/NN	Classical
Simulation Based	Maelstrom GENOM	Maelstrom	Delight.Spice Anaconda	FRIDGE	-	-
Equation based	-	SD-OPT ASTR/OBLX	Opasyn	AMGIE	-	Maulik GPCAD
Learning Based	Alpaydin	Alpaydin	GENOM	-	Alpaydin GENOM	-

A significant part of the tools initially employed simulated annealing but later SA was used more frequently as a complement to other techniques, i.e., ALPAYDIN [59], MAELSTROM [57]. Combinations of two or more different methods are named hybrid methods (sometimes the hybridization of EA with local search techniques is also known as Memetic algorithms (see Sect. 3.1.4) and were developed to take advantage from the potentials of each solution. The idea is to create a new algorithm with improved capacity to explore the promising regions of the search space. For example, in MAELSTROM [57] system, Krasnicki et. al. applied a "parallel recombinative simulated annealing (PRSA) method which combines multiple simulated annealing algorithms that run concurrently and share information via a genetic algorithm scheme. The same group of researchers developed another variation called ANACONDA [56] that introduces constraints variations in transistor devices which incorporate a genetic algorithm, coupled with a local "pattern search" technique. The FASY [10] system is a fuzzy-logic based synthesis tool with simulated annealing for coarse and gradient search for fine optimization. The fuzzy logic chooses a topology from a pre-defined library. The originality of this approach is the use of a NN model, built from data collected in optimization runs that is employed to update the fuzzy rules. The ALPAYDIN [59] system is an analog integrated circuit synthesis that computes the device sizing using neural-fuzzy performance models and user defined equations. The neural-fuzzy model is used to estimate some of the AC performance metrics. The remainder AC performance metrics are modeled by user specific equations. The performance model is built from a set of training data collected from SPICE simulations. This system incorporates the effect of process variations but has a drawback since new equations must be calculated by the user for each new topology.

In conclusion, the analog circuit synthesis is really a demanding task, which a unique optimization algorithm could hardly solve. The development of new methodologies and techniques should be explored to increase the efficiency of analog circuit design. The trends verified in this area show that the solution for some of the most important approaches lies on the integration of several methods to combine the best of each one, and on the employing of models to reduce computation times.

2.3.4 Other Characteristics

In what concerns to the tools scope (see Table 2.6), the simulation-based and artificial intelligent methods take the high scores because they can be normally applied to a broad range of analog circuits and, similarly, they modify the design capabilities of the system without too much overhead. However, their scope depends on the simulator model. The equation based-approach, when dealing with manual design equations, has short scope, however, if equations are derived by symbolic tools, a better incorporation of new design problems is possible, increasing the scope of the tool. Knowledge based-approaches are generally close tools, because they are limited to a reduced number of architecture topologies and design objectives.

Table 2.6. Scope characteristic

	Knowledge	Equation	Simulation	AI/NN
Scope of the tool	-	+/-	+	+

The “Encapsulate Details”, “Interactive design”, “Bookkeeping Facilities” and “Implementation language” issues were not subject to comparative analysis between methodologies once they do not depend on design methodology but result from the merit of each tool in particular.

2.3.5 Summary

The first efforts in the development of CAD tools started with low abstraction level implementations targeting primarily small systems. Large and complex systems were decomposed into small building blocks employing the expert knowledge. The variety of existing tools and techniques covering several aspects of analog design are summarized in Fig. 2.7. The first generation of design automation tools was driven to the optimization of design parameters, leaving to the designer the task of selecting an appropriate architecture.

Since then, different types of selection topologies evolved ranging from template approaches, to bottom-up and top-down topology generation approaches, executing simultaneously or independently from sizing activities.

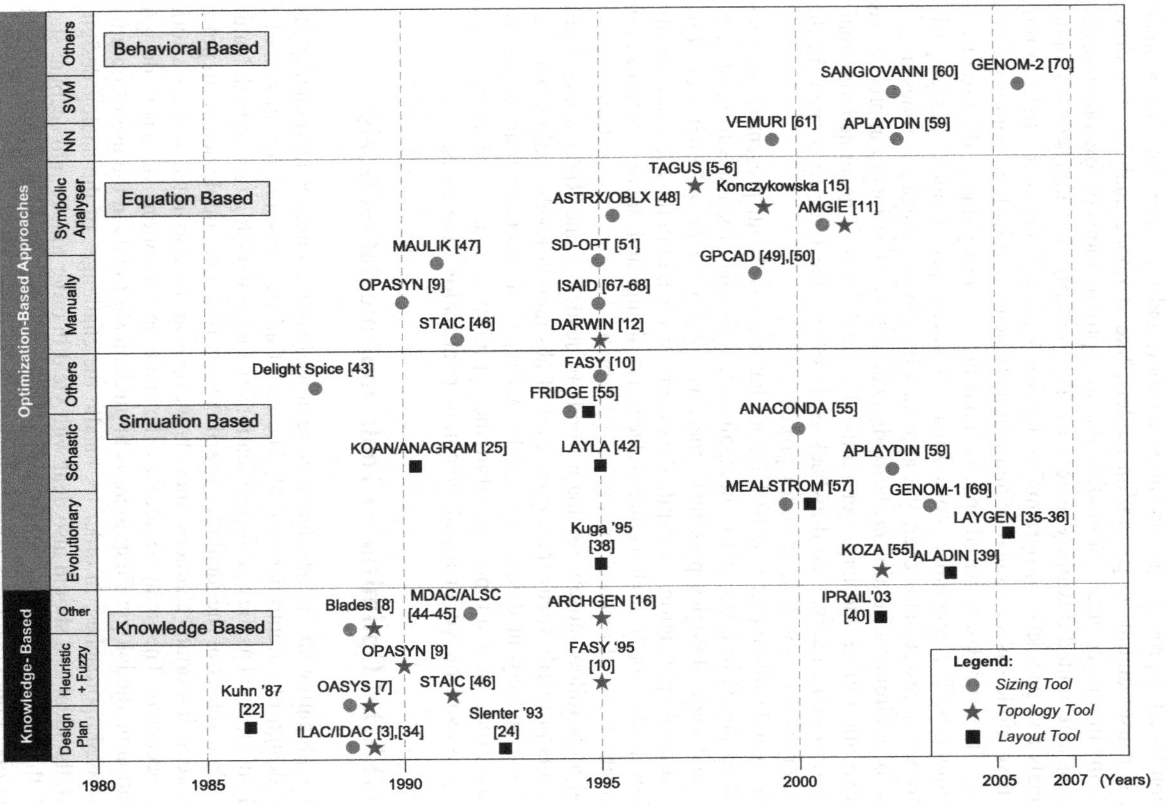

Fig. 2.7 Overview of analog synthesis tools

At the lower abstraction level, the knowledge methods based in special heuristic are out of date due to the long setup times involved, in the order of several weeks, which do not match the tight agenda of today's market pressure. In the equation based-approach, the run times are short and the setup can also be made short, if it uses automatic generation models, like symbolic analysis. The drawback is the limited accuracy of models, due to approximations or low-order design equations and limited flexibility in designs. The performance models based on posynomials and geometric programming foresee a great future if the time to produce these models is shortened or automatically generated without compromising accuracy. The simulation-based approach has high accuracy due to the use of circuit simulators. The generality is also high, allowing a large range of design problems to be addressed. However, the approach has longer execution times due to the use of a circuit simulator in the optimization loop. The model approach has short execution times and large generality. The model can be generated automatically and systematically. The drawback is, however, the large time spent in the preparatory phase as well as accuracy problems.

To conclude this topic, it must be said that effectively all presented methods have some points in favor and some against. Despite the broad spectrum of techniques and methodologies presented, there is not any "*defacto*" implementation for this area of applications. Despite the evolution verified in the high and low abstraction levels, both architecture selections, sizing and layout optimization remains the focus of research in analog EDA methodologies. The industrial commercial tools follow closely the main trends in academia and R&D workgroups. Their tools primarily focus the lower level of abstraction levels dealing with device sizing and layout description levels. All types of available frameworks assume the existence of a topology before the optimization run. Hence, no topology synthesis is available yet in any of the commercial analog EDA tools.

2.4 GENOM Optimization Tool: Implementation Goals

A tool aggregating all the best features, reviewed above, imposes hard challenges for the design of an automation tool. The set of all best covered features can be roughly interpreted as the main specifications of an ideal analog design tool. Naturally, only a subset of the ideal tool specifications is usually implemented in practice. Several important characteristics, however, can be appointed so that a tool can be accepted. They can be seen as the main specifications of a new design automation methodology. First, there is an undeniable trends in the use of optimization-based approaches, in order to, handle the challenges of the analog design. Second, the ideal tool should also deal with yield in order to take into account statistical fluctuations (process variations) inherent to the fabrication process and varying operating conditions (supply voltage or temperature variations), to make the design as robust as possible. Moreover, the design correctness and accuracy should be as close as possible to the industry electric validation tools. Furthermore, the overall optimization methodology should be as efficient as possible. Due to the existing trade-off between accuracy and computation time (section 2.3.2), this important goal can not be treated individually. However, the performance achieved by the resulting tool should outperform the traditional methods or existing methodologies or tools. In order to have wide acceptance this tool should allow the designer to modify the design configuration in a short time. A graphical user interface has to be supplied in order to increase the productivity. The GUI interface adds reporting information, as the designer is able to evaluate some dynamic parameters of the optimization process and carry out some configuration steps (interactive design, and flexibility). As finally, the resulting application should be preferentially designed in an independent platform or integrated with current EDA design environments. The interaction with externals tools should be carried out with open standards, if possible, to make the application integration in industrial design easier.

Following the trends presented by several modern tools, GENOM combines state-of-the-art modeling and searching techniques to deal with the complexity of analog circuit design problem. Since it cannot be granted that derivatives of the objective functions are known for the generality of this multiobjective problem, we have to trust non-derivative optimization methods, hence this thesis assumes these methods as the best choice. To ensure the design correctness and accuracy, GENOM employs a standard simulation tool in the loop of a modified genetic algorithm kernel allowing the corner simulations. To increase the efficiency of the evolutionary algorithm, a machine learning algorithm based on SVM was introduced. The proposed approach results in a new GA-SVM learning scheme applied to analog circuit design composed by the interaction of two machine learning engines. GENOM is primarily designed to increase the automation level and so it encapsulates much of intrinsic algorithm parameters from normal users but it permits some algorithm parameter changes to restricted users, through a configuration file. To allow a better use of available resources GENOM allows the execution either in a single processor machine or in a multiprocessor distributed environment. For efficiency reasons the GENOM code was written in C, therefore, the default user interface is text file oriented despite it has built-in functions which allow it to integrate a graphical in-house design environment. The following chapters will explore the details of this new tool.

2.5 Conclusions

Automated design of analog circuits, also referred to as analog circuit synthesis, has been the subject of active scientific research for many years now. This chapter has covered some of the most significant design automation methodologies applied to analog IC design automation. Here, a set of general properties, that allow the characterization of each approach, was been identified and a better insight related to advantages and limitations has been presented.

The characterization of each different approach supports the definition and identification of the general specifications for a new design automation methodology to be implemented in GENOM, a tool that will be applied to the automation of mixed-analog ICs.

References

[1] Horta, N.C.: Analog and mixed-Signal IC design automation: Synthesis and optimization overview. In: Proc. 5th Conference on Telecommunications, Tomar, Portugal (2005)

[2] Martens, E., Gielen, G.: Classification of analog synthesis tools based on their architecture selection mechanisms. Integration, the VLSI Journal 41(2), 238–252 (2007)

[3] Degrauwe, M., et al.: IDAC: An interactive design tool for analog CMOS circuits. IEEE J. Solid-State Circuits 22, 1106–1115 (1987)

[4] Lourenço, N.: LAYGEN: Automatic layout generation of analog ICs, from a system to device level using both hierarchical template descriptions and intelligent computing techniques. Master thesis, Dept. Electrical and Computer Engineering, Instituto Superior Técnico, Portugal (2007)

[5] Horta, N.C., Franca, J.E.: High-Level data conversion synthesis by symbolic methods. In: Proc. IEEE Int. Symposium on Circuits and Systems, vol. 4, pp. 802–805 (1996)

[6] Horta, N.C.: Analogue and mixed-signal systems topologies exploration using symbolic methods. In: Proc. Analog Integrated Circuits and Signal Processing, vol. 31(2), pp. 161–176 (2002)

[7] Harjani, R., Rutenbar, R.A., Carley, L.R.: OASYS: A framework for analog circuit synthesis. IEEE Trans. Computer-Aided Design 8, 1247–1265 (1989)

[8] El-Turky, F., Perry, E.: BLADES: An artificial intelligence approach to analog circuit design. IEEE Trans. Computer-Aided Design 8, 680–692 (1989)

[9] Koh, H.Y., Sequin, C.H., Gray, P.R.: OPASYN: A compiler for CMOS operational amplifiers. IEEE Trans. Computer-Aided Design 9(2), 113–125 (1990)

[10] Torralba, A., Chávez, J., Franquelo, L.G.: FASY: A fuzzy-logic based tool for analog synthesis. IEEE Trans. on Computer-Aided Design of Integrated Circuits and Systems 15(7), 705–715 (1996)

[11] Gielen, G., et al.: An analog module generator for mixed analog/digital ASIC design. Wiley Int. J. Circuit Theory Applications 23, 269–283 (1995)

[12] Kruiskamp, W., Leenaerts, D.: DARWIN: CMOS opamp synthesis by means of a genetic algorithm. In: Proc. ACM/IEEE Design Automation Conference, pp. 550–553 (1995)

[13] Stehr, G., Pronath, M., Schenkel, F., Graeb, H., Antreich, K.: Initial sizing of analog integrated circuits by centering within topology given implicit specifications. In: Proc. IEEE International Conference on Computer-Aided Design, pp. 241–246 (2003)

[14] Horta, N.C., Franca, J.E.: Algorithm-driven synthesis of data conversion architectures. IEEE Trans. Computer-Aided Design Integrated Circuits 16(10), 1116–1135 (1997)

[15] Konczykowska, A., Bon, M.: Structural synthesis and optimization of analog circuits symbolic analysis techniques. IEEE, Los Alamitos (1998)

[16] Antoa, B.A., Brodersen, A.J.: ARCHGEN: Automated synthesis of analog systems. IEEE Trans. VLSI Systems 3(2), 231–244 (1995)

[17] Lohn, J.D., Colombano, S.P.: A circuit representation technique for automated circuit design. IEEE Trans. Evolutionary Computation 3(3), 205–219 (1999)

[18] Koza, J.R., Bennett, F.H., Andre, D., Keane, M.A., Dunlap, F.: Automated synthesis of analog electrical circuits by means of genetic programming. IEEE Trans. Evolutionary Computation 1(2), 109–128 (1997)

[19] Sripramong, T., Toumazou, C.: The invention of CMOS amplifiers using genetic programming and current-flow analysis. IEEE Trans. Comput. Aided Design Integrated Circuits 21(11), 1237–1252 (2002)

[20] Leenaerts, D., Gielen, G., Rutenbar, R.A.: CAD solutions and outstanding challenges for mixed-signal and RF IC design. In: Proc. IEEE/ACM International Conference on Computer Aided Design, pp. 270–277 (2001)

[21] Aggarwal, V.: Analog circuit optimization using evolutionary algorithms and convex optimization. Master thesis, Dept. of Electrical Engineering and Computer Science, Massachusetts Institute of Technology (2007) `http://web.mit.edu/varun_ag/www/msthesis.pdf` (Accessed March 2009)

[22] Kuhn, J.: Analog module generators for silicon compilation. In: Proc. VLSI System Design, pp. 75–80 (1987)

[23] Wolf, M., Kleine, U., Hosticka, B.J.: A novel analog module generator environment. In: Proc. European Conference on Design and Test, pp. 388–392 (1996)

[24] Beenker, G., Conway, J., Schrooten, G., Slenter, A.: Analog CAD for consumer ICs. In: Huijsing, J., Plassche, R., Sansen, W. (eds.) Analog Circuit Design, pp. 347–367. Kluwer Academic Publishers, Norwell (1993)

[25] Cohn, J., Garrod, D., Rutenbar, R.A., Carley, L.R.: KOAN/ANAGRAM II: New tools for device-level analog placement and routing. IEEE J. Solid-State Circuits 26, 330–342 (1991)

[26] Carley, L., Georges, G., Rutenbar, R.A., Sansen, W.: Synthesis tools for mixed-signal ICs: Progress on frontend and backend strategies. In: Proc. Design Automation Conference, vol. 33, pp. 298–303 (1996)

[27] Lampaert, K., Gielen, G., Sansen, W.: A performance driven placement tool for analog integrated circuits. IEEE Journal of Solid-State Circuits 30, 773–780 (1995)

[28] Khademsameni, P., Syrzycki, M.: A tool for automated analog CMOS layout module generation and placement. In: Proc. IEEE Canadian Conference on Electrical and Computer Engineering, pp. 416–421 (2002)

[29] Yılmaz, E., Dündar, G.: New Layout generator for analog CMOS circuits. In: Proc. 18th European Conference on Circuit Theory and Design, pp. 36–39 (2007)

[30] Hartono, R., Jangkrajarng, N., Bhattacharya, S., Shi, C.: Automatic device layout generation for analog layout retargeting. In: Proc. International Conference on VLSI Design, vol. 36, pp. 457–462 (2005)

[31] Lampaert, K., Gielen, G., Sansen, W.: Analog layout generation for performance and manufacturability. Kluwer Academic Publishers, Dordrecht (1999)

[32] Jingnan, X., Vital, J., Horta, N.C.: A SKILLTM-based library for retargetable embedded analog cores. In: Proc. Design Automation and Test in Europe Conference and Exhibition, pp. 768–769 (2001)

[33] Jingnan, X., Serras, J., Oliveira, M., Belo, R., Bugalho, M., Vital, J., Horta, N.C., Franca, J.: IC design automation from circuit level optimization to retargetable layout. In: Proc. 8th IEEE International Conference on Electronics, Circuits and Systems, vol. 1, pp. 95–98 (2001)

[34] Rijmenants, I., Schwarz, Y.R., Litsios, J.B., Zinszner, R.: ILAC: An automated layout tool for CMOS circuits. IEEE Journal of Solid-State Circuits 24(2), 417–425 (1989)

[35] Lourenço, N., Horta, N.C.: LAYGEN – An evolutionary approach to automatic analog IC layout generation. In: Proc. IEEE Conf. on Electronics, Circuits and System, Tunisia (2005)

[36] Lourenço, N., Vianello, M., Guilherme, J., Horta, N.C.: LAYGEN – Automatic layout generation of analog ICs from hierarchical template descriptions. In: Proc. IEEE Ph. D. Research in Microelectronics and Electronics, pp. 213–216 (2006)

[37] Zhang, L., Kleine, U.: A genetic approach to analog module placement with simulated annealing. In: Proc. IEEE Int. Symposium on Circuits and Systems, vol. 1, pp. 345–348 (2002)

[38] Handa, K., Kuga, S.: Polycell placement for analog LSI chip designs by genetic algorithms and tabu search. In: Proc. IEEE Conference on Evolutionary Computation, vol. 2, pp. 716–721 (1995)

[39] Zhang, L., Kleine, U.: A novel analog layout synthesis tool. In: Proc. IEEE Int. Symposium on Circuits and Systems, vol. 5, pp. 101–104 (2004)

[40] Jangkrajarng, N., Bhattacharya, S., Hartono, R., Shi, C.J.: IPRAIL: Intellectual property reuse based analog IC layout automation. Integration, the VLSI Journal 36(4), 237–262 (2003)

[41] Castro-Lopez, R., Fernandez, F.V., Guerra-Vinuesa, O., Vazquez, A.: Reuse Based Methodologies and Tools in the Design of Analog and Mixed-Signal Integrated Circuits. Springer, Heidelberg (2003)

[42] Cory, W.: Layla: A VLSI Layout Language. In: Proc. 22nd ACM/IEEE Conference on Design Automation, pp. 245–251 (1985)

[43] Nye, W., Riley, D.C., Sangiovanni-Vincentelli, A., Tits, A.L.: DELIGHT.SPICE: An optimization-based system for the design of integrated circuits. IEEE Trans. Computer-Aided Design 7(4), 501–519 (1998)

[44] Leme, C., Horta, N.C., Franca, J.E., Yufera, A., Rueda, A., Huertas, J.L., et al.: Flexible silicon compilation of charge redistribution data conversion systems. In: Proc. IEEE Midwest Symposium on Circuits and Systems, pp. 403–406 (1991)

[45] Horta, N.C., Franca, J.E., Leme, C.A.: Framework for architecture synthesis of data conversion systems employing binary-weighted capacitor arrays. In: Proc. IEEE Int. Symposium on Circuits and Systems, pp. 1789–1792 (1991)

[46] Harvey, J.P., Elmasry, M.I., Leung, B.: STAIC: An interactive framework for synthesizing CMOS and BICMOS analog circuits. IEEE Trans. Computer-Aided Design 11(11), 1402–1417 (1992)

[47] Maulik, P.C., Carley, L.R.: Automating analog circuit design using constrained optimization techniques. In: Proc. IEEE Int. Conf. Computer-Aided Design, pp. 390–393 (1991)

[48] Ochotta, E.S., Rutenbar, R.A., Carley, L.R.: Synthesis of high-performance analog circuits in ASTRX/OBLX. IEEE Trans. Computer- Aided Design 15(3), 273–294 (1996)

[49] Hershenson, M., Boyd, S., Lee, T.: GPCAD: A tool for CMOS op-amp synthesis. In: Proc. IEEE/ACM Int. Conf. Computer-Aided Design, pp. 296–303 (1998)

[50] Hershenson, M.M., Boyd, S.P., Lee, T.H.: Optimal design of a CMOS Op-Amp via geometric programming. IEEE Trans. Computer-Aided Design 20(1), 1–21 (2001)

[51] Medeiro, F., Verdu, B.P., Vazquez, A.R., Huertas, J.L.: A vertically integrated tool for automated design of modulators. IEEE Journal of Solid-State Circuits 30(7) (1995)

[52] Gielen, G., Wambacq, P., Sansen, W.: Symbolic analysis methods and applications for analog circuits: A tutorial overview. Proc. IEEE 82, 287–304 (1994)

[53] Wambacq, P., Fernandez, F.V., Gielen, G., Sansen, W., Rodriguez-Vazquez, A.: Efficient symbolic computation of approximated small signal characteristics. IEEE J. Solid-State Circuits 30, 327–330 (1995)

[54] Daems, W., Gielen, G., Sansen, W.: An efficient optimization–based technique to generate posynomial performance models for analog integrated circuits. In: Proc. 39th Design Automation Conference, pp. 431–436 (2002)

[55] Medeiro, F., et al.: A Statistical optimization-based approach for automated sizing of analog cells. In: Proc. ACM/IEEE Int. Conf. Computer-Aided Design, pp. 594–597 (1994)

[56] Phelps, R., Krasnicki, M., Rutenbar, R.A., Carley, L.R., Hellums, J.: ANACONDA: Simulation-based synthesis of analog circuits via stochastic pattern search. IEEE Trans. Computer-Aided Design of Integrated Circuits and Systems 19(6), 703–717 (2000)

[57] Krasnicki, M., Phelps, R., Rutenbar, R.A., Carley, L.R.: MAELSTROM: Efficient simulation-based synthesis for custom analog cells. In: Proc. ACM/IEEE Design Automation Conference, pp. 945–950 (1999)

[58] Liu, H., Singhee, A., Rutenbar, R.A., Carley, L.: Remembrance of circuits past: Macromodeling by data mining in large analog design spaces. In: Proc. Design Automation Conference, pp. 437–442 (2002)

[59] Alpaydin, G., Balkir, S., Dundar, G.: An evolutionary approach to automatic synthesis of high-performance analog integrated circuits. IEEE Trans on Evol. Computation 7(3), 240–252 (2003)

[60] Bernardinis, F., Jordan, M.I., Sangiovanni-Vincentelli, A.: Support vector machines for analog circuit performance representation. In: Proc. Design Automation Conference, pp. 964–969 (2003)

[61] Wolfe, G.A.: Performance macro-modeling techniques for fast analog circuit synthesis. Ph.D. dissertation, Dept. of Electrical and Computer Engineering and Computer Science, College of Engineering, University of Cincinnati, USA (1999)

[62] Synopsys Inc.: Products and solutions-HSIM, PowerMill, NanoSim (2009), `http://www.synopsys.com` (Accessed March 2009)

[63] EEDesign, Synopsys acquires ADA for analog boost (2009), `http://www.eetimes.com/` (Accessed March 2009)

[64] Synopsys Inc.: Circuit Explorer - analysis, optimization & trade-off (2009), `http://www.synopsys.com` (Accessed March 2009)

[65] Cadence Inc, Products: Composer, Virtuoso, DIVA, NeoCircuit, NeoCell, UltraSim, NcSim (2009), `http://www.cadence.com` (Accessed March 2009)

[66] Toumazou, C., Makris, C.: Analog IC design automation: Part I - Automated circuit generation: New concepts and methods. IEEE Trans. Computer-Aided Design 14, 218–238 (1995)

[67] Makris, C., Toumazou, C.: ISAID: Qualitative reasoning and trade-off analysis in analog IC design automation. In: Proc. IEEE Int. Symposium on Circuits and Systems, pp. 2364–2367 (1992)

[68] Toumazou, C., Makris, C.A., Berrah, C.M.: ISAID - a methodology for automated analog IC design. In: Proc. IEEE Int. Symposium on Circuits and Systems, vol. 1, pp. 531–535 (1990)

[69] Barros, M., Neves, G., Guilherme, J., Horta, N.C.: Enhanced genetic algorithm kernel applied to a circuit-level optimization E-Design environment. In: Proc. 10th IEEE International Conference on Electronics, Circuits and Systems, pp. 1046–1049 (2003)

[70] Barros, M., Neves, G., Guilherme, J., Horta, N.C.: Analog Circuits Optimization based on Evolutionary Computation Techniques. Integration, the VLSI Journal 43(1), 136–155 (2010)

3 Evolutionary Analog IC Design Optimization

Abstract. This chapter starts with an overview on computation techniques aiming to solve nonlinear optimization problems with emphasis on evolutionary optimization algorithms and discusses their relevance to analog design problem. The main virtues and weaknesses, as well as, the design issues of evolutionary algorithms are discussed with a description of the recent developments in this field. This chapter also introduces a new optimization kernel based on genetic algorithms applied to analog circuit optimization. It includes a detailed description of the coding schemes, the fitness function, the genetic operators and other design strategy criteria. Finally, a robust IC design methodology supported by the optimization kernel is presented in the end of the chapter.

3.1 Computation Techniques for Analog IC Design – An Overview

During the past decades significant activities have been carried out on the analog design automation focusing the problem of automatically sizing the circuit, automating topology selection and layout generation. The following sections review some computation techniques used to solve the analog IC design problem.

3.1.1 Analog IC Design Problem Formulation

Analog IC design has been seen as the hard topic of IC design for a long time. As discussed in the introductory chapter the main reason for this design effort is that analog design is knowledge-intensive, due to the deeply nonlinear behavior of the performance measures and the strong sensitivity of these measures to variations in the design parameters, having as result design problems with extremely complex trade-offs. Mathematically, the analog and mixed design problem (AMDP) can be formulated through the following general nonlinear programming (NP) expression [1]-[5] of a general multi-objective problem:

$$\begin{gathered} Optimize\,\mathrm{F}(\vec{x}) \\ subject\,to: \Omega=\left\{\vec{x}\in\Re^{n} \mid \mathrm{G}(\vec{x})\le 0\right\} \end{gathered} \tag{3.1}$$

M.F.M. Barros et al.: Analog Circuits and Systems Optimization, SCI 294, pp. 49–88.
springerlink.com

where, $\vec{x}$ is multidimensional vector of decision parameters in $\Re^n$ delimited by an upper and a lower bound given by $x^i_{min} \le x^i \le x^i_{max}$. $F(\vec{x})$ represents the vector of *m* objectives ($f_1(\vec{x}),..., f_m(\vec{x})$) to be minimized or maximized and $G(\vec{x})$ the vector of *p* constraints that must be satisfied to guarantee the feasibility of the solution. When *m* equals to one, the expression 3.1 corresponds to a single objective problem, when *m* is greater than one, it stands for a multi-objective problem. Normally, the elements of $G(\vec{x})$ are handled explicitly by inequality expressions relating the desired value of hard specifications taking form $g_i(x) \le Specs_i$ or $g_j(x) \ge Specs_j$ or $g_k(x) = Specs_k$ with $i + j + k = p$. In the branch of operational research (OR), the equality constraints can alternatively be transformed into a pair of inequality constraints taken from $|g_k(x)| - \varepsilon \le Specs_k$, where ε is a small allowed tolerance. From now on, this transformation will be implicitly assumed every time inequality constraints are referred in the text. The domain space Ω is a nonempty set in $\Re^n$ and the objective functions are $f_i : \Re^n \to \Re$.

The estimation of each design alternative, concerning one or several different objective functions and multiple constraints represent a global, high-dimensional optimization problem. The aim of this multi-objective, multi-constraint problem is to catch up the best relation between circuit performance (e.g. power dissipation, circuit area, gain, gbw, etc.,) and design parameters (e.g. the width W, length L, resister values R, capacitor values C,...) subject to some constraints (e.g. geometry constraints, designer rules, etc). Specifically, the undertaken design problem tries to find the particular values of the design parameters ($x_1^*, x_2^*,..., x_n^*$) belonging to Ω which yield a point or a region in the performance space of the objective functions that satisfies the required specifications as illustrated in Fig. 3.1.

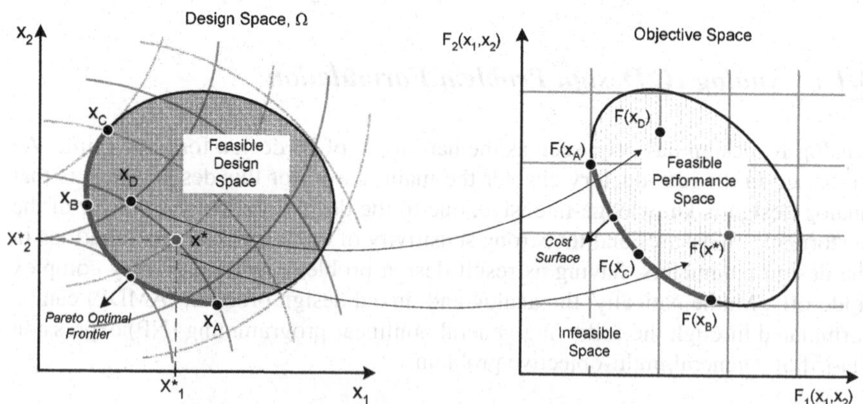

Fig. 3.1 Basic concepts in multi-objective optimization

From this point forward any abstract task accomplished to solve a problem or to look for the best solution can be perceived as a search through a space of potential solutions. A number of approaches have been described in the last chapter to find the global optimum of the cost surface associated to a high-dimensional optimization (AMDP) problem. These approaches can be classified as knowledge-based methods, employing design knowledge and heuristics, and optimization-based methods, making use of numerical programming techniques. When seeking for the decisive objective (ultimate goal) i.e., finding the global optimum solution with a minimum number of function evaluations or running time, both approaches present some strengths and limitations. The work developed in this research is committed to explore an optimization-based approach. The next sub-section briefly reviews some of the most promising algorithm techniques to solve such a complex problem and explains the choice taken to structure the framework presented in this book.

3.1.2 Numeric Programming Techniques

Both the research community and the industry have been paying extra attention to optimization algorithms for the past few years. Optimization algorithms have become an important research area due to their efficiency in achieving approximate solutions to NP-hard [3] problems and in solving problems where no analytic method applies, for instance, solving nonlinear differential equations. Besides, optimization algorithms can be applied to a wide range of situations, as most scientific and industrial design problems may be formulated through an optimization task whose aim is to minimize or maximize a given objective function and might involve linear or nonlinear constraints, integer and/or continuous variables, stochastic or deterministic inputs, and single or multiple criteria objectives. In the field of numeric optimization there is a vast range of optimization methods, the most of which can be categorized according to [6] in Fig. 3.2.

Some of these methods may be better adapted to the nature of the specific problems pointed out by the intermediate nodes or applied to a wide spectrum of problems like stochastic programming. Therefore, the knowledge of the problem nature allows the choice of more suitable optimization algorithms. Picking the correct optimization algorithm is an essential step to obtain the best trade-off between accuracy and time efficiency in all optimization problems. For example, when the problem is exclusively represented by a set of linear equations, linear programming (LP) techniques are more appropriated. The simplex algorithm, developed by George Dantzig [7]-[8], is one of the most popular methods to solve LP problems, like the “traveling salesman” which aims to find the minimal traveling distance. For this kind of problems, equality constraints are welcome, because it is known that if the optimum exists, it is situated at the surface of the convex set, whereas inequalities can be manipulated mathematically to equalities by the addition of slack variables [3].

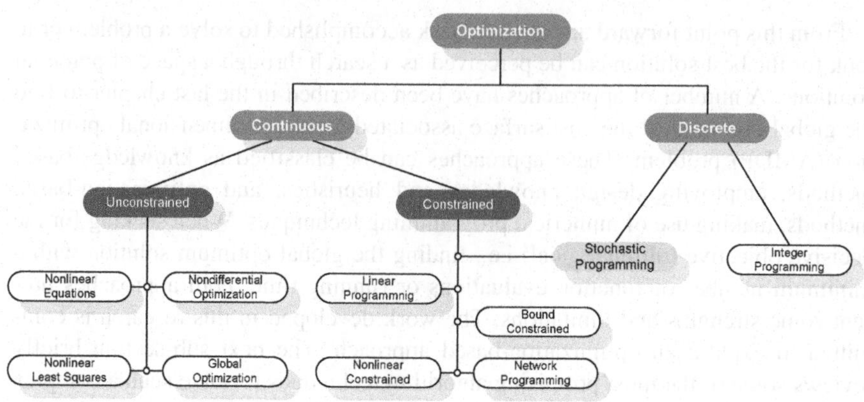

Fig. 3.2 The optimization tree outlines of the major algorithms in each area

However, determining the global optimum to the general nonlinear programming problem can be a challenging task since there is no specific method capable to do it in a systematic way [3], although it can be found in certain circumstances, i.e., when the objective functions and the constraints satisfy certain properties. Appendix B will briefly address the characteristics of some general purpose optimization techniques for nonlinear problems including random search methods (Appendix B.1), gradient-based methods (Appendix B.2), constraints programming (Appendix B.3), stochastic methods (Appendix B.4), and multiple objective optimizations (Appendix B.5). The table 3.1 briefly resumes the described optimization methods as well as their main advantages and limitations.

3.1.3 The No-Free-Lunch Theorem

To look for the best performance algorithm in the field of optimization algorithms is considered a utopia. A very important theorem generally accepted by the community, known as the "*No Free Lunch Theorem*" (*NFL*) [23] states that it cannot exist any algorithm which solves all kinds of problems. On average the "performance of any pair of algorithms across all possible problems is identical". A possible illustration of this theorem can be seen in the Fig. 3.3 where on average both algorithms perform equally well when considering a broad range of different problems. In particular, if some algorithm *A* outperforms *B* over some set of optimization problems, then the reverse must be true over the set of all other optimization problems.

Although there is a long list of available optimization algorithms it is not known any method tailored to deal directly with the complexity of Analog Design problem in order to obtain the best trade-off between performance accuracy and time efficiency. A great part of them are customized to some specific class of

Table 3.1 Properties of general purpose optimization methods for nonlinear problems

Optimization method	Type of problems & description	Advantages	Limitations
Random search	Global unconstrained. Consists in selecting randomly potential solutions and evaluating them.	The easiest form of heuristic search. Often used as a reference tool. One known example is the Monte Carlo (MC) method.	Blind search, doesn't use any domain-specific information to guide the search; search is usually slow.
Gradient based [9]	Local unconstrained nonlinear optimization which applies the concept of successive search, based on the information of gradient or derivative function.	Used for local search, improved version like Newton method converges fast.	Requires the derivative of objective function and uni-modal spaces. Only takes into account local information.
Optimization method	Type of problems & description	Advantages	Limitations
Constraint Programming [10]-[13]	Constraints continuous or discrete. Penalizes the solutions that are near or violate the constraints boundaries with an amount proportional to constraint violation.	Models complex problems easily. Increase the efficiency of the search using the constraints to prune the search space. Mature tools.	Weakness when dealing with cyclic dependencies.
Stochastic Search [14]-[16]	Local and global search continuous or discrete. Does not require a continuous, a convex or differentiable cost function.	Model Multi-Objective, multimodal, Multi Constraints, Nonlinear Objectives. Encloses a High spectrum of applications.	Performance efficiency. Because of their probabilistic nature the global optima requires many iterations to converge.
Multi-objective Opt. [17]-[19] [20]-[22]	Global search continuous or discrete. Problems requiring the optimization of more than one conflicting objective functions.	Model Multi-Objective, Multi Constraints, Nonlinear Objectives, Trade-offs	Performance efficiency.

problems exploiting certain features and accordingly to the NFL theorem they are able to achieve high performance patterns. In the classical optimization techniques, the majority of the proposed methods are predominately local in scope, relies on derivatives and are not robust enough in discontinuous, vast multimodal or noisy search spaces [25], so they are more efficient in solving linear, quadratic, strongly convex, unimodal and many other special problems. On the other hand, stochastic algorithms and specially the evolutionary algorithms own a set of intrinsic properties (reviewed in preceding section) which allow them to deal with highly complex optimization problems, like the analog IC design – defined as a non-lineal, high dimensional, high constrained, multi-objective problem. However, it cannot be expected that an optimizer based on stochastic algorithms could give good results for all spectrum of applications (no free lunch). Stochastic optimizers are not considered a black box system, hence they must be tailored with expertise for each specific problem. This will be the subject of the next section where an analog design optimization tool matching the analog IC design problems (for the design of IC circuits) based on a standard stochastic evolutionary algorithm will be presented in detail.

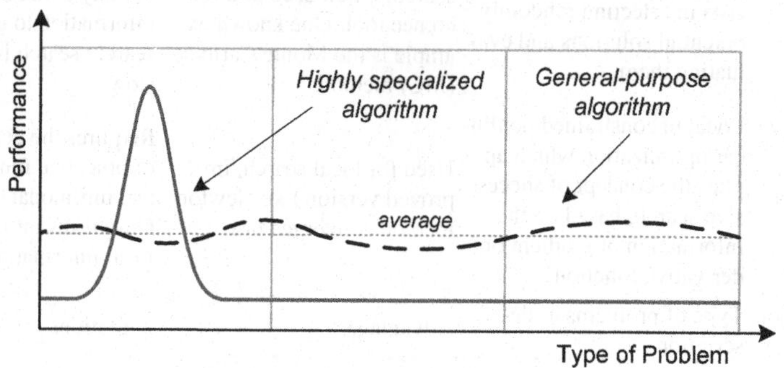

Fig. 3.3 The no-free-lunch theorem representation [24]

3.1.4 Evolutionary Computation Techniques Overview

In the past few years, evolutionary computation (*EC*) [24]-[27] has gained increasing notability since it is becoming the method of choice for solving complex problems especially when classic methods cannot be efficiently applied or have a difficult formalization [28]. Besides the advantages inherited from stochastic algorithms, EAs own several characteristics that make the difference from other optimization and problem solving techniques [29]. Table 3.2 summarizes its main characteristics. In few words, evolutionary computation constitutes a class of iterative and stochastic optimization techniques inspired by concepts from Darwinian natural evolution theory, namely the genetic inheritance and the strife for survival.

Evolutionary computation embraces a range of programming techniques such as genetic algorithms [30] [32][33][34][35], evolution strategies [31], evolutionary programming [32]-[33] and genetic programming [34]-[35]. Evolution Strategies (*ES*) and evolutionary programming (*EP*) were developed independently at the same time as genetic algorithms (*GA*). Although these techniques have the same aims and use the same basic structure cycle, there are slight differences related to the representation of candidate solutions and the implementation of selection, recombination and mutation operators. The Table 3.3 resumes the described techniques. At present there are no big differences between these approaches. Many of the algorithms only differ in slight details, because of the constant interchange and crossing of ideas between the different approaches. As far as representation and type of operators is concerned, most researchers came to the decision that the best solution representation should be achieved according to each specific problem.

Table 3.2 EAs main characteristics

Properties	Description
Flexible	They can adapt easily to different types of problems or can be applied in a problem with little prior knowledge, avoiding in-depth mathematical representation which is difficult and sometimes impossible to acquire for some complex problems.
Simple	They allow short timings for model setup and easy changes of the problem.
Robust	They can be effective in noisy environments.
Adaptive	They can deal with self-adaptation, allowing dynamic changes of process parameters.
Decentralized	Due to the ability to lead with populations of solutions, they are easily parallelizable [36], [26], taking advantage of the power of distributed and higly parallel computing environments.

Table 3.3 Milestones in Evolutionary Techniques - Overview

Evolutionary Techniques	Main Contributions	Activity Period
Evolution Strategies by Ingo Rechenberg and Hans-Paul Schwefel [31]	Introduce the continuous parameter optimization and expand the mutation operator to continuous stochastic variations. Don't use crossover operator. A new concept of breeding based on $(\mu+\lambda)$-ES and (μ,λ)-ES strategies. Use self-adaptation to adjust control parameters of the search.	In the 1960s and early 1970s
Evolutionary Programming by Fogel [32],[33]	Applies the FSM concept to represent candidate solutions and use variation and selection strategies adapted to this environment. Evolves populations of solutions with mutation and selection.	At the end of 1960s
Genetic Algorithms by Holland's (Original) [30]	Use discrete encoding representation, traditionally in binary as strings of 0s and 1s. Apply the simple evolutionary algorithm in optimization problem able to evolve toward better solutions.	Became popular in 1970s
Genetic Programming by Koza [34],[35].	Represents individuals as executable hierarchical trees of computer programs (code) that can be mutated by changing or swapping subtrees representing many different kinds of problems.	Begining of the 90'

During the last few years there has been significant progress in evolutionary computation techniques and the field of applications has expanded considerably. There are some other approaches which adopt mechanisms from nature. Table 3.4 reviews the most important trends in this domain and summarizes the main advantages and disadvantages of recent EC techniques.

Table 3.4 Recent trends in evolutionary computation - Overview

Optimization Algorithms	Main Contributions	Drawbacks
Ant Colony Optimization (ACO) [37] was introduced in and early 1990s	Deals with the parallel search and use of memory structures to hold information on the quality of the historical results. They have an advantage over other stochastic algorithms like SA and GA. When the graph changes dynamically; the ant colony algorithm may adapt to changes in real time.	Oriented for solving hard combinatorial and constraint discrete optimization problems. Coding is not straightforward.
Particle Swarm Optimization (PSO) [38] was introduced in middle of 1990s	Is conceptually simple due to a small number of parameters to adjust and is oriented for parallelization. It does not require many user-defined parameters. Is flexible because it can be designed for local minimization as well as allows the incorporation of algorithms or heuristics for global optimization.	Parameters dependency. Slow convergence in the vicinity of the global optima.
Estimation of Distribution Algorithms (EDA) [39],[41]	Incorporates methods for automated learning between variables. Uses probabilistic models considering discrete or continuous, independent or dependent variables. The crossover and mutation operations were replaced by estimation and sampling of a probability distribution. In some application outperforms GAs.	EDAs are not efficient or applicable to the continuous optimization, real-time optimization and multi-objective optimization.
Differential Evolution (DE) [42] was introduced in middle of 1990s	Easy to use method based on EAs. The variations schemes implemented in DE to create offspring, automatically execute a step size adaptation as the search process converges toward good solutions.	There are a number of variations (schemes) and it is unclear which scheme performs the best under static conditions.
Cultural Algorithms (CA) and Immune Systems [43]	The knowledge is the fundamental key to achieve the requirements of a decision making process. Apply techniques in order to acquire knowledge and save them in the "belief space" and then use it to bias the search. This technique was used in GENOCOP tool.	This technique only deals with linear constraints, as the original GENOCOP.

There are some other approaches involving hybrid systems. Classical simple EAs usually cannot compete with other state-of-the-art algorithms that are specifically adapted to some particular type of problems. On the other hand, the demand for even more accurate and efficient evolutionary algorithms in a broad range of applications led to the development of many hybrid approaches, where an evolutionary algorithm can be combined with local search heuristics and problem-specific variation operators or expert encodings. These hybrid approaches also known as *memetic algorithms* [44]-[46], employ several *metaheuristics* such as *simulated annealing*, *tabu search* and *guided local search* methods in combination with EAs, in order to efficiently improve the exploration process of major areas of the search space, as well as, the local exploitation related to the fine-tuning of the most promising candidate solutions [47]. There are other approaches that combine EAs with exact optimization techniques [48] such as, *dynamic programming*, *branch-and-bound*, and *integer linear programming* techniques. Very often these hybrid approaches extended by the problem of specific knowledge, outperform the standard evolutionary methods, as well as, other standard techniques.

Nowadays, many of the state-of-the-art EA-based techniques are rather complex, problem-specific hybrid systems. EAs can be described as very flexible tools since they are able to be hybridized with problem specific techniques to improve performance. In conclusion, combining an EA-framework with other techniques should be recognized as a contribution of great worth.

3.2 Key Issues in Evolutionary Search

EAs have a rich historical background of experience and research, oriented for the optimization of the convergence processes that consistently finds an approximate solution quickly and efficiently, suitable for a broad range of applications.

The basic evolutionary process described in Appendix C.1 and exemplified in Appendix C.1.3 for the optimization of constrained problems, contains a minimal set of features that make evolutionary algorithms competitive to solve hard global optimization problems. They provide a set of unique properties that allow dealing with a broad range of nonlinear problems where traditional optimization techniques, like gradient descent, hill climbing, and purely random search, are often inefficient or inadequate. Due to its success and usefulness already proved in engineering applications, the branch of evolutionary computation has been object of continuous development. Table 3.5 lists of the main ongoing research themes related to evolutionary optimization found in the most recent literature.

Table 3.5 Key issues in EAs

	Description	Advantages and Disadvantages
Constraint Handling [[11],[13]]	In the original form, EAs do not define a mechanism able to guide efficiently the search towards the feasible region in constrained search spaces. A wide variety of techniques have been adopted to handle all kind of constraints such as the use of, penalty functions, specialized representation and operators, repair mechanisms, separations of objective and constraints and hybrid methods. From the universe of EA the most common approach is the use of penalty functions to those solutions that violate constraints. The most common approach uses the amount of constraint violation to penalize an infeasible solution, thus promoting the selection of feasible solutions.	In spite of the great variety of methods there is not a proved method to solve all different sort of constraint (linear, nonlinear, etc) problems [44]. This means that the suitable chosen method when there is no knowledge about the domain is still an open research problem. Since the penalty based approaches are easy to implement and are also quite efficient, they are often used as the first choice in spite of their known limitations. Penalty functions require a precise judgment of the penalty factors so that the right combination of penalties is discovered, thus a balance between feasible and infeasible solutions will be met. A disadvantage of this approach is that it is dependent on the problem.
Population Diversity [[49],[51]]	Any evolutionary optimization technique requires a mechanism in order to maintain diversity in the population. If there is no diversity, the search will be concentrated only in one area of the feasible region, in a phenomenon known as genetic drift. There are several ways to maintain diversity, among them, niching methods and the use of mutation have become the most popular ones. The niching methods are the extension of EAs and make it possible to find more than one local optimum of a function. Sharing and crowding are the best known and popular niching techniques. Both aim to decrease the fitness of individuals that are located in crowded regions in order to promote the proliferation of solutions in sparse regions. Sharing uses the concept of distance to its closer neighbor while crowding is controlled by the density of solutions in a region or population.	The use of mutation methods to increase diversity is one of the main goal in EAs (highlight in 3.3.5). However, there is not any deterministic formula concerning the optimal settings wherever it is used in static or dynamic mutation mode. When comparing niching approaches, the sharing method presents several drawbacks. Sharing has difficulty in distinguishing the local optima that are much closer to each other than the niche radius. Thus it is necessary to know a priori the distribution of the optimum and define the suitable value of niche radius. Special attention should be given to evolutionary operators, for example, the mating between chromosomes in different niches may often produce unsuccessful offspring.

Table 3.5. *(continued)*

	Description	Advantages and Disadvantages
Speed of Convergence [44]-[46]	The convergence speed in evolutionary optimization follows three independent branches of research. One explores the mixture of complementary optimization techniques to improve the performance of the overall algorithms. The hybridization of local search techniques with EAs known as memetic algorithms can also improve the convergence speed near the optimal. Another approach explores the multiple solutions available in distributed computation. A different approach uses dynamic reduced models or approximate models (will be focused later) to accelerate genetic algorithm based on design optimization.	There is a great variety of hybrid methods that combine the best of each technique [57-59]. Although no mature methodology has been established yet, research has proved the efficiency of these approaches to increase the speed and sometimes the accuracy in a variety of problems. The increase of efficiency using distributed computation techniques is an obvious attractive approach to take advantage of the increasing capacity of computation resources available nowadays. EAs coding structures are relatively simple to adapt to distributed environment.
Multi-objective multi-constraint problems [17]-[19],[4][5],[20]-[22],[55]-[57]	The area of research known as Evolutionary Multi-Objective Optimization, or EMOO for short, is one subject of constant active research in field of evolutionary computation. EMMOs are designed with regard to two common goals, obtaining a fast and reliable convergence to the Pareto front and ensuring a good distribution of solutions along the front. The main themes of research are focused on techniques for handling constraints, maintain diversity of the solutions, hybridization with other local search methods and archiving for storing non-dominated vectors. The main influent approaches are MOGA, NSGA, NPGA and SPEA as highlight in [4]-[5],[19]-[22].	The main themes of research are similar to general EAs, however they differ in methodology. MOO is tailored to deal with multi-objective optimization problems. A great debate exists around the quality and the virtues of SOO and EMOO approaches. Recent studies [61] which compare the performance of a single-objective genetic algorithm with an EMOO approach indicate that there is no dominant method when comparing the performance in set of multi-objective problems. The results obtained demonstrate that EMOO algorithm outperforms SOO in some cases but does not work well on problems with many objective optimization functions.

Table 3.5 (*continued*)

	Description	**Advantages and Disadvantages**
Evolutionary algorithms with high computational cost evaluations [47], [58]	Despite the great success of the use of Evolutionary algorithms in some industrial and engineering design problems, it may become impractical in the case of high computational cost evaluations or when an explicit objective function does not exist. One approach to overcome this problem is to estimate the objective function by constructing approximate (or surrogate) models that are used to replace exact but expensive evaluations, thus reducing the design computational cost. Several models have been used, ranging from response surface models (or polynomials), the krigging model, very popular in design and analysis of computer experiments (DACE), neural networks, support vector machines, etc. The use of approximate models in evolutionary computation can also lead to a reduction in true objective function calls. A detailed survey of fitness approximation in evolutionary computation can be found in [48].	The use of cheap surrogate models to be used in lieu of exact models makes EAs a viable tecnology to be applied to computationaly expensive problems. Aproximate models are in order of magnitude cheaper to run. However, the integration of approximated models with EAs to real-world problems have met limited success. The main drawback is that the computational cost demonstrated specially with response surface models (which involve low-order polynomial regression) as well as with the krigging model becomes unacceptable as the dimensional of the problem increases. This effect known as the "curse of dimensionality", represents the amount of complexity caused by the exponential increase in the solution space with the addition of a extra dimension.
Adaptive parameters [59][60]	Adaptive parameters is the branch in research of EAs dealing with the control parameter or strategy parameters that allow the adjustment of parameter values during the evolutionary process. Parameters controlled during the run have gained much attention and their techniques inspired from ESs domain can be classified in three main groups: deterministic, adaptive and self-adaptive control mechanisms. In the deterministic control, the strategy parameters are modified by some deterministic rules but without using the feedback from the search. The adaptive approach uses the knowledge adquired in evolutionary process to refine the strategic parameters. In the self-adaptive approach, no direct feedback control is used, the control parameters are treated as optimization variables encoded in the chromossome which evolves during the evolution process using the standard algorithms operators.	In spite of the increase of flexibility when compared with the static parameter tuning mechanisms, the deterministic an adaptive approaches require in both cases, the right definition of deterministic rules which could be difficult to obtain. The adaptive with feedback control has an additional advantage over the deterministic since the feedback returned from the process may help to decide if the trend with the new parameter value should persist or not. The use of self-adaptive techniques simplifies the problem. In the mutation control parameter case the optimized parameter is related with the speed of step size adaptation rather than the step sizes themselves. Their default values can be applied with success for certain types of problems even though it is not guaranteed the fastest adaptation scores.

3.3 GENOM - Evolutionary Kernel for Analog IC Design Optimization

Here a new approach to multi-objective optimization, GENOM, is introduced. Fig. 3.4 overviews the main building blocks of GENOM optimization system. The modified GA kernel which forms the central unit of this system is surrounded by the evaluation engine unit and other additional units with computation facilities involving remote communications and data processing. This section discusses the internal structure of the analog optimization tool, which was extended with new concepts, from an initial standard genetic algorithm to a modified GA implementation (GA-MOD) leading to an improvement in both robustness and efficiency. The proposed methodology corresponds to a simulation based approach, since it can be applied to all types of design circuits, producing highly accurate results and providing an extended layer of analysis, concerning the robust design required in the industrial environment.

Next, the optimization algorithm kernel, as well as, its particular aspects will be described, such as the multi-objective multi-constraint function formulation, the structure representation, the evolutionary control strategy, the self-adaptive parameters, premature convergence, etc.

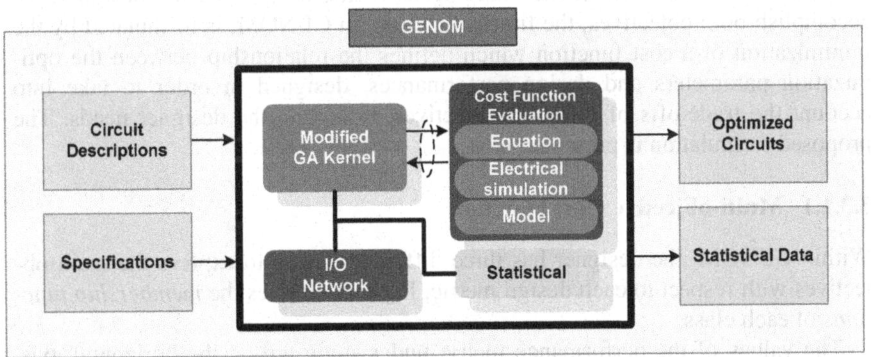

Fig. 3.4 GENOM system overview

3.3.1 Fitness Function Study

The optimization methods introduced new challenges when solving multiobjective problems. Designers have to formulate a fitness function that better represents the objectives of the problem, and need to setup the decision maker (DM) preferences in the presence of multiple conflicting design metrics. DM preferences, defined before, during or after the optimization process [13],[58]-[61], express the importance level of each objective and can take the form "high", "medium" or "low".

DM preferences incorporated in the fitness function will assist the selection of the optimal solution, providing that one exists.

One of the easiest and perhaps most widely used method to carry out performances trade-offs is the weighted sum approach (3.2). DM preferences are taken into account by assigning several weightings, for each objective function *fi(x)*. A weighted sum approach transforms a multi-objective optimization problem in a single-objective optimization problem.

$$\min \sum_{i=1}^{p} w_i * f_i(x)$$
$$s.t.\ \mathrm{x} \in \mathrm{S}, w_i \geq 0, \forall \mathrm{i} = 1,\ldots,\mathrm{p} \text{ and } \sum_{i=1}^{p} w_i = 1 \qquad (3.2)$$

In spite of the simple formulation, there is no definite articulation between the weightings and the obtained solution. The determination of the weightings from the decision makers' preferences is not an accurate procedure, either. This method presents another disadvantage, as with convex combination of different objectives, solutions at non-convex part of the pareto-front cannot be located [61].

Both user preferences and fitness function are the key factors for the effectiveness of the optimization problem. The fitness function of a multi-objective optimization problem must reflect the exact needs of the design and the designer. To accomplish both objectives, the fitness function, in GENOM, is formulated by the minimization of a cost function which defines the relationship between the optimization parameters and design performances, designed in order to take into account the trade-offs of different objectives, reflecting the designer needs. The proposed formulation is presented next.

3.3.1.1 Multi-objective Cost Function

Within GENOM, the designer has three different classes to express the main objectives with respect to each design metric. Fig. 3.5 outlines the *membership functions* of each class.

The values of the performance metric under study are on the horizontal axis, and the class-function to be minimized for that design metric, is on the vertical axis. Each class, representing a unique desired behavior, is available in two versions, the *soft* and the *hard constraints*. The soft class functions are aggregated in an objective function whose goal is to find the preferred solution among some preference criterion and the hard classes become the constraints because they represent the absolute limitations imposed on the system.

The above framework, derived from the Physical Programming methodology [62],[63], presents a more flexible and user-friendly solution as it grasps the designer's physical conception of the aimed design. In this method designers give a desired value which is employed to form a class function, instead of providing weights to establish priority on objectives. These class functions aim at the convergence of the algorithm and at the same time provide a complete set of *pareto*

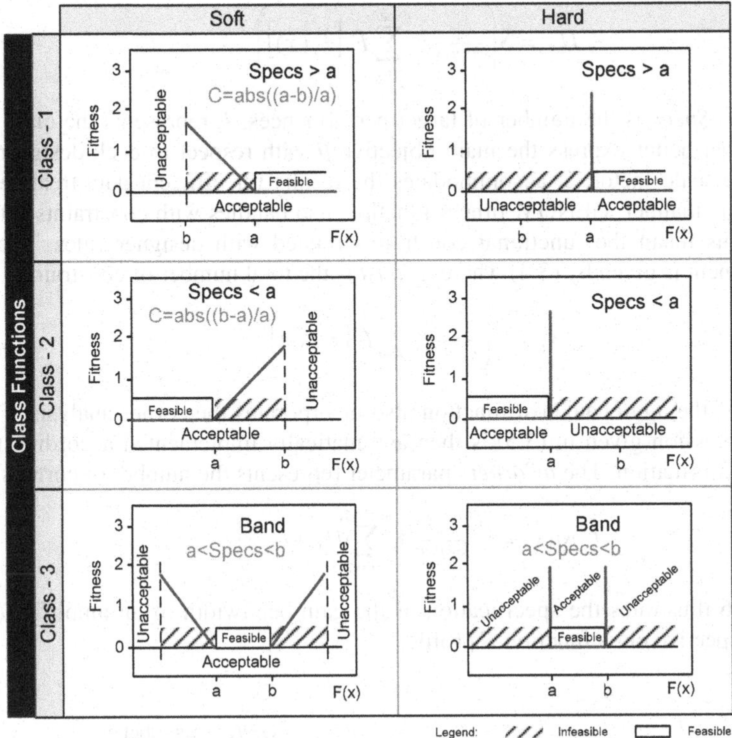

Fig. 3.5 Classification of preferences for each performance metric

points. Like in *fuzzy logic*, the given set of values help quantifying an acceptable or unacceptable result as they fit within some tolerance into the desired objective. The tolerance boundaries are automatically determined accordingly to the target values previously defined for a given performance metric. One advantage of this method is that the designer does not need to provide information related to weightings and only has to concentrate on data concerned with circuit design. When compared to strict preferences, *fuzzy preferences* like the ones depicted in Fig. 3.5 have another advantage. They can improve the quality of solutions in the evolutionary cycle. For example, if an infeasible sample is very close to the feasible region, missing almost all specs and using a strict preference, it will be, almost certainly, discarded from the evolutionary cycle due to a hard penalizing factor. However, the employment of *fuzzy preferences* together with an appropriate selection method, as used in this approach, greatly increases the likelihood of the survival of this sample, and so, its potential schemata will have a chance to evolve.

In GENOM, the fitness function was formulated as the sum of several *aggregated cost functions*. The aggregated cost function presented in the following expression (3.3) measures the design specs satisfaction degree.

$$U_S = S\left(\frac{1}{nSpecs} * \sum_{j=1}^{nSpecs} F\left[f_j(\vec{x})\right]\right) \tag{3.3}$$

Where *nSpecs* is the number of target performances, *F* represents the class functions that better express the main objective *fi* with respect to each design metric and the scale factor, *S*, accommodates the results for efficient data treatment. In the same manner, GENOM fitness function also handles with constraints satisfaction, this mean the functional constraints related with designer rules. The new component is given by (3.4) where *nConst* is the total number of constraints:

$$U_C = S\left(\frac{1}{nConst} * \sum_{k=1}^{nConst} F\left[f_k(\vec{x})\right]\right) \tag{3.4}$$

Finally, the aggregate cost function also incorporates the corner analysis. The final expression given in (3.5) is then automatically implemented according to the given information. The *nCorners* parameter represents the number of corners.

$$Cost(\vec{x}) = S\left(\frac{1}{nCorners} * \sum_{i=1}^{nCorners} (U_S + U_C)\right) \tag{3.5}$$

Fig. 3.6 illustrates the specification of the gain-bandwidth of an amplifier (gbw) and respective class function *F* profile.

Nº	Specs	Target	Units
1	GBW	> 200	MHz
2	DC gain	> 80	dB
3	Phase Margin	> 60	º

Nº	Extreme Specs	Limits	Units
1	GBW	< 125	MHz
2	DC gain	< 60	dB
3	Phase Margin	< 55	º

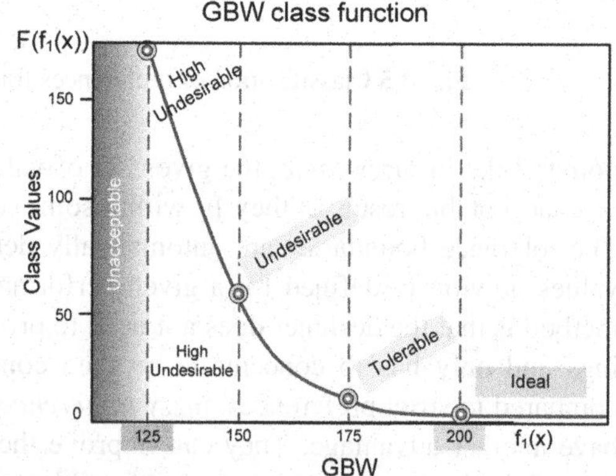

Fig. 3.6 Tolerance limits of a class function

This example adopts the *soft class-1* function that maximizes the *gbw* criterion. Between the desirable target specification and the other extreme admissible value, five regions that characterize the degree of desirability are created such as, the ideal, tolerable, undesirable, high undesirable and unacceptable. These regions are defined in the order of decreasing preferences.

With exception of *class-3*, four boundary values were necessary to build this class function. The band *class-3* function in general is defined by eight values. These values translate the designer preferences for each range of each given design metric. Only the two extreme values introduced by the designer are mandatory, the other ones can be automatically calculated. Whatever the case, the value returned by the specification class-function is the same at each of the region boundaries, regardless of class-type or criterion. Class functions are built in such way that the vertical excursion $F\,[f_i(\vec{x})]$ over two distinct criteria will always have the same vertical magnitude as long as one travels across the same region-type, even if the location of the boundary values ($f_i(\vec{x})$) changes from criterion to criterion. This property encapsulates a normalizing function in which every region-type is conditioned in the same way for different criteria.

3.3.1.2 Cost Function with No Preference Articulation

When it is not known any preference information about performance metrics beyond the target performances values, a weighted cost function (3.6) is being used giving each evaluation a satisfying degree (*rank*) of a candidate solution related to the desired specification. It privileges solutions with maximum satisfied specifications and distinguishes the best solution, as the one that minimizes the difference between achieved performance values and specified target value.

$$Cost(x) = \min \sum_{i=1}^{\substack{N^{o}\\Corners}} wc_i * f_i(x)$$

$$f_i(x) = \sum_{j}^{\substack{N^{o}\\Specif}} wp_j * \left|Fit_j(x)\right| + \sum_{k}^{\substack{N^{o}\\Const}} wc_k * \left|Ctr_k(x)\right| \qquad (3.6)$$

$$Fit_j(x) = \left(\frac{pj - goal(j)}{goal(j)}\right) \text{ and } Ctr_k(x) = \left(\frac{ck - goal(k)}{goal(k)}\right)$$

$Fit_j(x)$ is a set of normalized objective functions derived from goals and performance specifications to be optimized and *$Ctr_k(x)$* is a set of normalized user-defined functional constraints. Once the user specifies an upper or a lower bound for the design constraints (goal), these are used to translate the achieved design and constraints specifications in cost function profiles accordingly to the Fig. 3.5. The indices *j* and *k* are the number of objective and functional constraint functions, respectively. The aim of these normalized functions is to assign an equal importance to each competing specification. The designer can now setup the relative importance for each competing specification adjusting the individual scalar weights wp_j and wc_k. At the beginning, greater weight values can be assigned to important design objectives and design constraints. With this formulation the optimized algorithm can further explore the solution space and generate more than one circuit feasible solution.

Table 3.6 Normalization overview

n	Type	SPECs	TARGET	Achieved Performance (pj)	Objective Function $Fit_j(x)$	$f_i(x) = \sum_{j}^{N^o Specif} wp_j * \lvert Fit_j(x) \rvert$
1	Perf.	DC gain	Gain > 80	90 dB	0	$f_i(x) = 0$
2	Perf.	GBW	Gbw > 200	150 MHz	-0.25	$f_i(x) = 0.25$
3	Goal	Power	**min (power)**	**← Valid spec.**	**(*)**	**(*)**
4	Goal	Power	min (power,0,10)	5 mw	0.5/10	$f_i(x) = 0.25+0.05$

(*) Problem dependent.

Table 3.6 gives an overview of the two different types of normalization applied in GENOM.

The no.1 performance specification was entirely fulfilled so the respective class objective function is null and the contribution to the aggregate cost function $f_i(x)$ is zero. The no.2 performance specification is the same type as no.1, however, the target was not achieved, so the respective class objective function $Fit_j(x)$ produces a value proportional to the missing target value. This type of normalization aims to balance the intensities of all performances, as well as, all function constraints.

A particular case is devoted to the design objective of the problem. In GENOM the *design objectives* or *goals* specs can be defined between two intervals (minimum and maximum) as defined with spec no.4. Alternatively spec no.3 can also be used however, the *maximum* and *minimum* is calculated automatically as described in Fig. 3.7. The normalization for design goals specs follows expression (3.7). The contribution for the final aggregate cost function is scaled down by a factor of 10, $Fit_j(x) = \frac{Fit_j(x)}{10}$, called the *residual*. This procedure prevents the appearance of a dominant individual with extremely lower goals that does not fulfill one or more regular performance specs.

$$Fit_j(x) = \begin{cases} \frac{|pj - min|}{(max - min)} & ,(min \le pj \le max) \vee (pj < min) \\ 1 + \frac{|pj - max|}{(max - min)} & ,others \end{cases} \tag{3.7}$$

The very first idea behind GENOM fitness formulation is, first, to seek for the feasible solutions (satisfying all *Perf.* constraints) and after, optimize its goals (power, area, etc.). Once a feasible solution is found, the aggregated fitness value is composed only by the sum of all *goal residuals* since the fitness related to performance and *functional constraints* is zero by definition. The remaining non-zero residuals will be used in order to optimize the goals of the problem and at the same time, provide a set of *pareto* points whose weights are automatically adjusted accordingly to the goals, as shown in Fig. 3.7.

```
IF (Stop Condition = First Solution) EXIT execution as soon as 1st solution is found,
ELSEIF (goals type= 3) {
    - Calculate the mean and standard deviation of the respective goal spec, m and s.
    - Maximum = m+Bs; Minimum = [0,.., m-Bs]          /*B is integer between 1..5 */
(a) IF (PARETO) {
    - remGen= maxGen-actualGen;                /* remaining number of generations */
    - Share = (remGen/numObjectives);      /* during this period minimize one goal */
    FOR (each goal) {
        Use weights to optimize one goal and relax the others;
        Each optimization evolves during "Share" generations;
        /*This promotes the sampling towards the extreme of pareto front*/
    }
  }
(b) OTHERWISE, optimize the design objectives without using weights
  - all wp_j=1 in cost function until the end of generations.             /*default*/
```

Fig. 3.7 Optimization of design objectives and pareto weights management

3.3.2 Individual Encoding, Population Structure and Sampling

The optimization algorithm is built over a single population structure that allows the existence of elite individuals, following the (μ+λ) steady-state model [31],[33],[64]. In the (μ+λ)-ES, a population of μ individuals produce λ offspring per generation and the selection process reduces all individuals (μ+λ) to just μ individuals again. The present population structure is divided into four main specific regions as illustrated in the Fig. 3.8. Only a fraction of the population reproduces and dies each generation. The proportion of each region is in conformity with generally accepted EAs practices [56]-[57],[65].

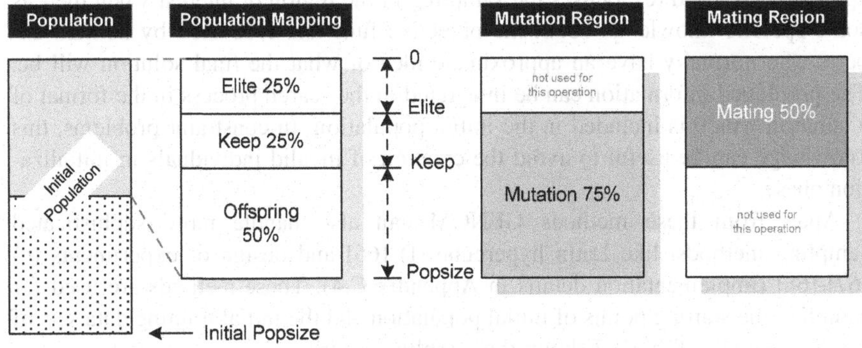

Fig. 3.8 Population structure

Each individual in the population is represented by a chromosome of real-coded values, illustrated in Fig. 3.9, because it is considered the representation that best matches the primary target related to continuous domain applications.

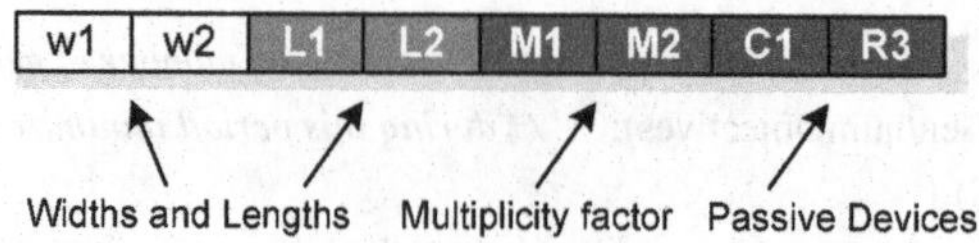

Fig. 3.9 Chromosome type is a vector of real numbers

In order to achieve a better coverage of the search space, the initial population in GENOM is created by sampling the search with the double of individuals in the population size. The estimation of the population size is based on a heuristic rule which involves the estimated search space size and is restricted to a value between 32 and 128 individuals.

In EAs there is no recommended or imposed initialization method to be followed. The default method in GENOM implements a random initialization following a uniform distribution as follows:

$$x_{i,j}^{(0)} = x_{j\min} + (x_{j\max} - x_{j\min})u_j \quad i \in \{1,...\lambda\}, j \in \{1,...,n\} \tag{3.8}$$

where,

u_j is a random number uniformly distributed over [0; 1],
$x_{i,j}$ denotes the *j-th* component of a vector x_i and λ denotes the population size.

The aim is to create a population with a good coverage of the search space, in order to find the regions of most promising solutions. Another variation of this approach is to impose a regular grid-layout where the sampling points are evenly divided all over the space, as illustrated in Fig. 3.10. Another variation of these approaches (d) and (e) focuses the sampling in the region of interest when there is some specific knowledge about the objective function. It is used by domain experts who normally have an approximate idea of what the final solution will be. The privileged information can be integrated in the search process in the format of a solution which is included in the initial population. In constraint problems, this knowledge can be useful to avoid the creation of invalid individuals in initialization phase.

Apart from these methods GENOM can also handle more sophisticated sampling methods, like Latin hypercube (f) [66] and design of experiments (c) [67]-[68] (implementation details in Appendix C.4). These methods are used for sampling the starting points of initial population and the initial training data sets of learning models. Table 3.7 shows the sampling criteria.

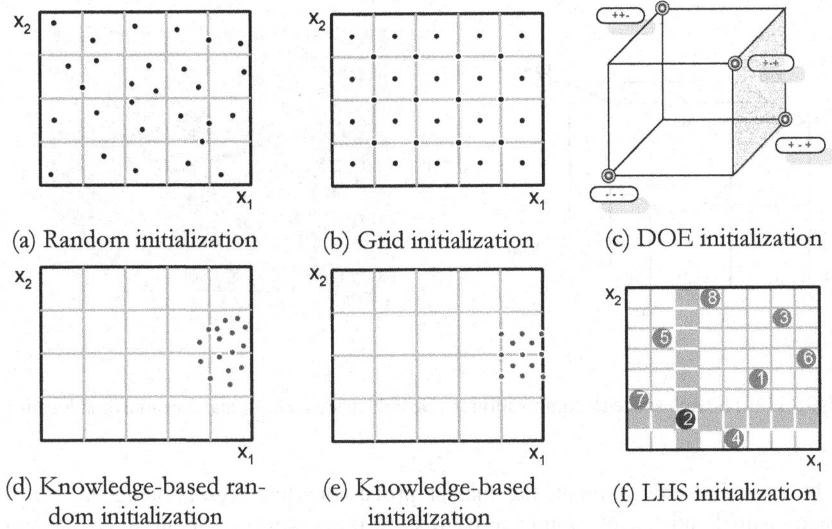

(a) Random initialization (b) Grid initialization (c) DOE initialization

(d) Knowledge-based random initialization (e) Knowledge-based initialization (f) LHS initialization

Fig. 3.10 Sampling strategies

Table 3.7 Sampling criteria

Methods	**Description**	**Criteria**
Random	Purely random initialization	Optional
Grid	Used to build models and display surface plots	Optional
Knowledge-based Grid & random	There is some specific knowledge about the problem	Optional
DOE (Appendix C.4)	Optional. Applied after a feasible region is found.	Optional
LHS (Appendix C.4)	Default	Default

Besides that, the search space decomposition (SD) was introduced, in order to reduce problem complexity and the number of cost function evaluations, therefore improving GA efficiency. Basically, it consists of a divide-to-conquer strategy which decomposes the search space in subspaces, useful in a distributed environment. The search space decomposition consists of dividing each variable range or a subset of the variables in p parts, thus welding, at most, p^n problem subspaces, where n is the number of optimization variables, as illustrated in Fig. 3.11.

Fig. 3.11 (a) illustrates the particular case of three optimization variables, x, y, and z, considering a subdivision of each variable range in 2 parts. Fig. 3.11(b) illustrates an example of search space decomposition with non-trivial mathematical functions including the initial and final chromosome locations for the test with search space decomposition, where each subspace solution is located over a white spot, i.e., a low cost area.

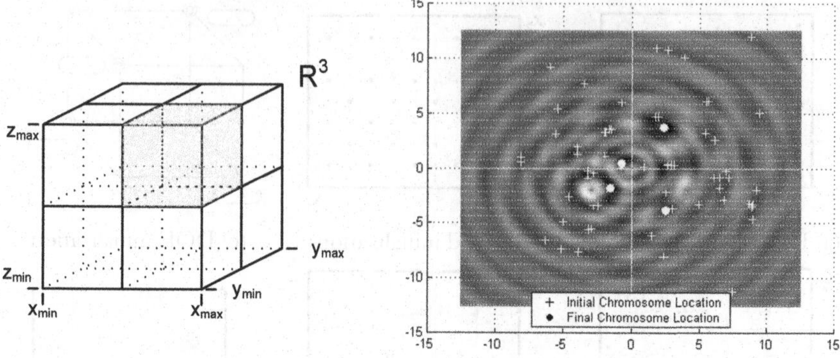

Fig. 3.11 (a) Search space decomposition (b) Search space contour and chromosome location

In a parallel environment, the master processor after decomposing the search space in small subspaces assigns to each slave processor the execution of one subspace optimization task. The slave task, executes one independent sequential GA of a unique search subspace. Then, the best chromosomes, from each slave processor run, are transferred to the global optimization array in the Master processor. Finally, when all the optimization sub-tasks are completed and the global optimization array is full with the best overall chromosomes from all search space, the Master processor executes a final global optimization task having those chromosomes as the initial population.

Table 3.8 illustrates the achieved performance measures, for a test with a non-trivial mathematical function presented in [69], executed for 100 runs with a maximum of 500 iterations each. This test also includes the algorithm modification introduced by *premature convergence prevention* defined in Sect. 3.3.5. There is an increase on efficiency, when using parallel processing, with the asymptotic limit of *1/n* CPU time compared to the serial processing approach, where *n* is the number of used processors.

Table 3.8 GENOM performance measures

Type of Runs	**Average no. of iterations**	**Average no. of cost func. evaluations**	**Average Minimum**
Standard GA	383	9265	-1.3377
w/ space decomposition(sd)*	367	9042	-1.3806
w/ premature convergence prevention (pcp)	175	4282	-1.4150
w/ sd and pcp*	174	4342	-1.4196

*Approximately 1/n CPU time when using parallel processing.

3.3.3 Selection Strategies

GENOM is based on a steady-state selection approach implementing an elitism (or *truncation*) strategy. The default selection mechanism employs a hybrid methodology between linear ranking and the tournament selection. First individuals are sorted according to a ranking algorithm with two levels of feasibility defined below, then the tournament selection will select within the current population, the parents that will create the next generation of individuals.

3.3.3.1 Ranking-Based Scheme

The GENOM selection algorithm uses *tournament selection* with a tournament size of two, preceded with a *feasibility-based sort algorithm*, inspired in K. Deb [11] and C. Coellho [12] settled in the following conditions:

(a) Both solutions are feasible;
(b) Both solutions are infeasible, or
(c) One solution is feasible but the other is infeasible.

Fig. 3.12 Deb's nuclear conditions

The new *feasibility-based sort* algorithm has the ability to make pair-wise comparison following the order rules of Fig. 3.13. The variant implemented in GENOM, begins the ranking process giving priority to individuals that better meet the feasibility region.

This method makes a separation of performance constraints and functional constraints, as described in 4.2.1, and compares the feasibility status of each solution in order to provide the search direction towards the promising (feasible) region based mostly on feasibility information (number of feasibility constraints satisfied) rather than in the constraint function value. When both individuals are feasible (satisfy all mandatory functional constraints), a similar process is followed in order to provide the search direction towards the promising performance region. This approach promotes infeasible solutions in the surroundings of the feasible region based in the number of constraints satisfied.

(a) IF both individuals k_1, k_2 are feasible (functional)
IF both satisfy the same number of the objective functions,
SELECT individual with the better value of the objective function;
ELSE select individual with the greatest number of the objective function satisfied;
(b) IF only one individual IS feasible(functional), SELECT it; and
(c) IF both individuals are infeasible,
SELECT the individual with smaller number of violated constraints,
OTHERWISE, IF the number of constraints satisfied IS equal,
SELECT the one with the smallest value of violated constraints.

Fig. 3.13 The GENOM tournament with feasibility-based ranking algorithm

The novelty of this approach is the ability to handle constraint-based problems which do not require any penalty parameter, so the problems which require penalty terms can be eliminated. Several similar variants were implemented in GENOM, like for example a ranking process giving priority to individuals that better meet the feasibility function, or another one that first provides the search direction towards the performance region and after that, towards the most feasible region. However, for analog problems with a higher number of feasibility constraints than design constraints the tournament with feasibility-based sort algorithm described above works better. The default method, tournament with feasibility-based sort algorithm, proves to be more effective in the generality of the experiments; in the same way as Deb conclude with his original work [11]. Optionally, the *roulette wheel* and *stochastic universal sampling* can be used as alternatives to tournament selection.

3.3.3.2 Constraint-Based Selection

A new alternative selection mechanism based on *feasibility knowledge constraints* is also introduced and it can be generalized to all types of problems. This method differs from the last selection schemes which were based on a probabilistic method, to a new one based on *knowledge* "*satisfiability*". The GENOM evaluation module is able to produce individual constraint information for each specific gene, notifying which constraints are satisfied or not and the respective amount of constraints violation. This knowledge, returned from the evolutionary process in the form of binary vectors called "*masks*" (see Fig. 3.15), is now available to other modules of optimization tools, like the *sort* and *pairing* routines responsible for selection process. The nuclear steps of the new constraint-selection scheme are explained in Fig. 3.14 and Fig. 3.15. The idea is to couple a pair of chromosomes with the largest number of genes that complement (or fulfill) the "missing" genes.

1. Sort by Feasibility method from population from
   0 – popsize -> (Rank-based Selection)
2. Use tournament selection with/out replacement 80% -> 0 - Keep
3. Use Selection by Matching Masks in 20%
4. Crowding- clusters similar chromosomes by randomly choosing a small number of chromosomes (3) and replacing the most similar, in terms of the distance. Avoid the repetition of pairs.

Fig. 3.14 The GENOM Constraint-based Selection scheme

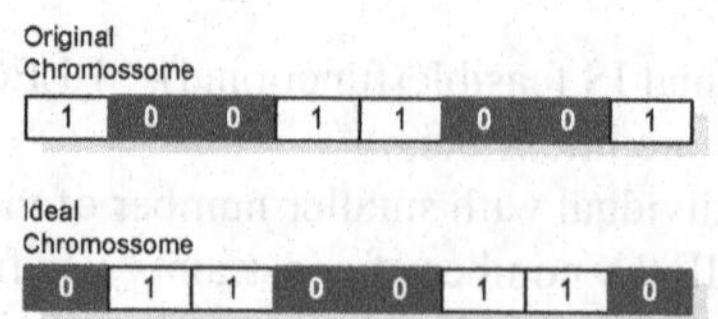

Fig. 3.15 Ideal pair using satisfiability constraints (masks genes)

In practice, the original pair is looking for the pair that XORed with itself giving the largest number of ones (a feasible solution).

For the pairing strategy, the selection of the potential mate is chosen from the list of candidates that better complement the already selected parent in terms of satisfied constraints. Mating the parent with the one that better fulfills the faulty constraints of both parents can potentially increase the probability of achieving a child that satisfies more constraints than the parents do.

3.3.4 *Crossover Strategies*

The basic crossover function implemented in GENOM uses the *Gaussian mutation* together with a *uniform crossover* to produce offspring solutions. This process is a mixture of the standard Gaussian mutation operator with a standard uniform crossover, as illustrated in Fig. 3.16.

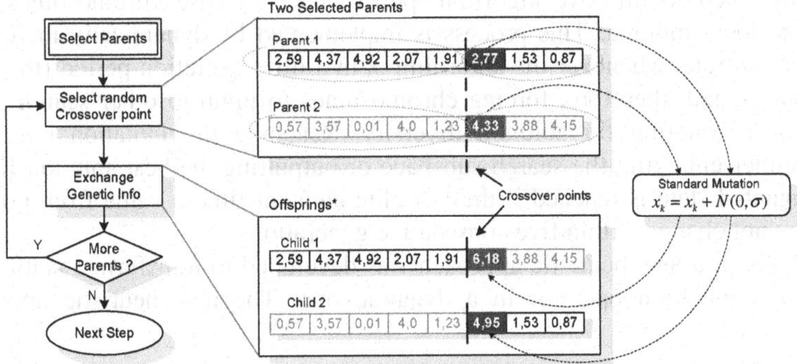

Fig. 3.16 The crossover operator with standard mutation

Another option is the use of standard arithmetic crossover with one weight λ as defined in Fig. 3.17 (b) as, $x_1' = \lambda . x_1 + (1-\lambda) . x_2$ or using a variation with a specific weight for each gene or in k (k<N) random genes xi in the chromosome $x' = (x_1, x_2, ..., x_n)$ defined as:

$$x_i' = \lambda_i x_{1i} + (1-\lambda_i) x_{2i}$$

$$x' = \sum_{j=1}^{k} \lambda_j x_j, \quad where\ \lambda_j \in [0,1], \quad \sum_{j=1}^{k} \lambda_j = 1 \tag{3.9}$$

The two first approaches of Fig. 3.17 (a) and (b) are used in GENOM for exploratory purposes while the arithmetic crossover with *n* weights is employed in the final stage of the evolutionary process or when a significant number of feasible solutions is found.

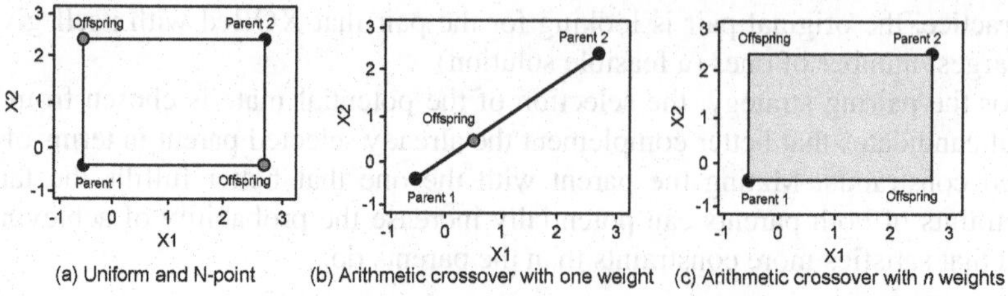

Fig. 3.17 Crossover for real chromosomes on a 2-D dimensional problems

3.3.5 Mutation Strategies

Mutation is one of the primary methods of maintaining diversity among feasible solutions. The basic mutation functions implemented in GENOM use the *standard uniform* and the *Gaussian mutation* according to the *mutation rate* previously defined. Additionally, a premature convergence prevention (PCP) process (Fig. 3.18) was introduced to improve algorithm performance in case chromosomes converged to local minima. This process is implemented by dynamically increasing the mutation rate, whenever the algorithm is in a little evolution period (the stagnated state), and, therefore, forcing chromosomes to jump to other search space locations, accounting for solution diversity. Reaching the mutation rate limit means either enlarging the search subspace or outputting the best solution found. The stagnated state is reached if the last elite element fitness value from population does not change within five consecutive generations.

In addition, a new heuristic approach was developed to transform a static control of the mutation operator in a dynamic one. The new heuristic approach

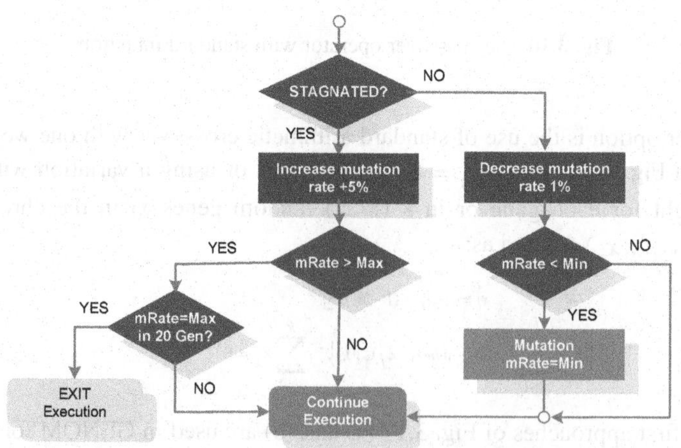

Fig. 3.18 GA w/ premature convergence prevention flowchart

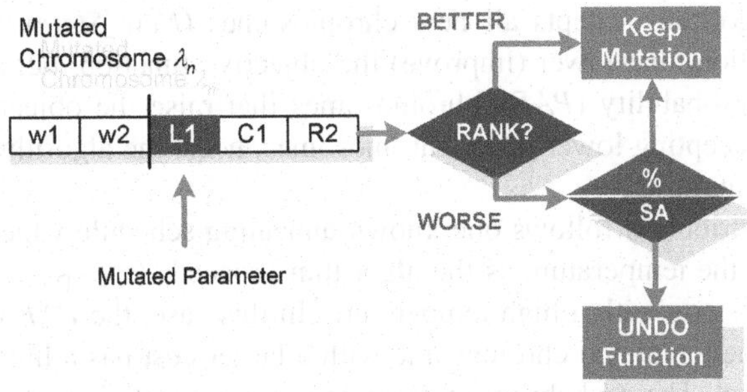

Fig. 3.19 Heuristics associated to the mutation operator

illustrated in Fig. 3.19, which can be optionally work together with PCP, applies an *UNDO* function to the mutate chromosome after verifying that the modification introduced does not get a better result.

The desired effect is to allow a better exploration of the search space in the beginning of the process and a better exploitation at the end, mimicking the principles of a *Simulated Annealing* (SA) algorithm (Appendix C.4). By accepting points of higher objective function (lower rank), the algorithm avoids being stacked in local minima, allowing a global exploration of the search space.

The UNDO function is ruled by a SA like algorithm following the *acceptance function* (PAF) and the *annealing temperature* (Temp) described in Fig. 3.20.

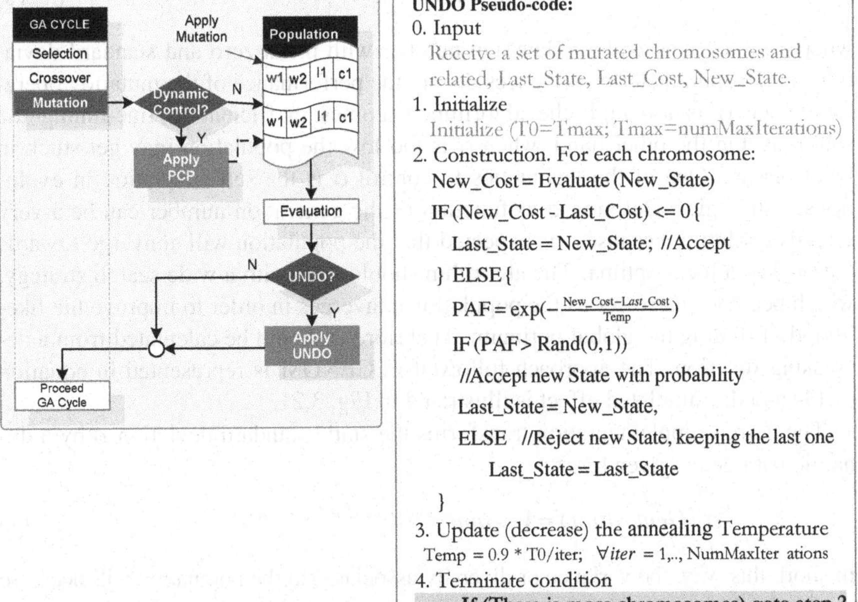

Fig. 3.20 Mutation control flow and code

Briefly, the algorithm accepts all new chromosomes (*New_State*), modified by mutation operation, that lower (improve) the objective function, but also, accepts with a certain probability (*PAF*), chromosomes that raise the objective function (larger cost). Accepting lower ranks chromosomes, helps the algorithms to escape from local minimum.

The *PAF* distribution follows one known annealing schedule which systematically decreases the temperature as the algorithm proceeds ($Temp = 0.9 * T0/iter$). The simulation starts with a high temperature. In this case, the *PAF* is very close to 1. Hence, a new mutated chromosome with a larger cost has a high probability of being accepted. The probability of accepting a worse state is high at the beginning and decreases at the temperature decreases. When T is high it promotes the *exploration* of search space, when it is low, the *exploitation*.

In this work, premature convergence prevention is the default method of maintaining diversity among the feasible solutions.

3.3.6 Step Size Control – Dynamic Evolutionary Control

In order to efficiently control the population diversity and progressively reduce the search space to the solution boundary, the optimization kernel implements the *Gaussian Mutation* together with *Gaussian Crossover* [57] to dynamically control the probability distribution applied when generating the offspring solutions. For the mutation operator case, each component x_i of vector x is replaced by x'_i ,

$$x_i^{'} = x_i + N(0,\sigma) \quad (3.10)$$

where, N(0,σ) is a random Gaussian number with mean zero and standard deviations σ. The parameter σ influences deeply the performance of the mutation operator. When σ is too high the algorithm becomes inefficient to fine-tuning the solutions. On the other hand, when σ is too low the population may get stuck in local optima. One of the techniques to control σ is the self-adaptation in evolutionary strategies. Defining σ as function of the generation number can be a very effective solution because it is expected that the population will converge towards a global or a local optima. The algorithm should start with a wide search strategy, which becomes narrower as the population converges in order to improve the likelihood of finding the global optimum. Therefore σ should be calculated from a decreasing function. The approach followed in GENOM is represented in equation (3.11) and the simulated effect is illustrated in Fig. 3.21.

The current implementation transforms the static standard deviation σ by a dynamic parameter, given by σ':

$$\sigma'(Generation) = 1 - rand^{(1-Generation/\,Max\_Generations)} \quad (3.11)$$

In short, this way, the variance or diversity associated to the population will decrease automatically as generations converge to an upper limit (*Max_Generations*).

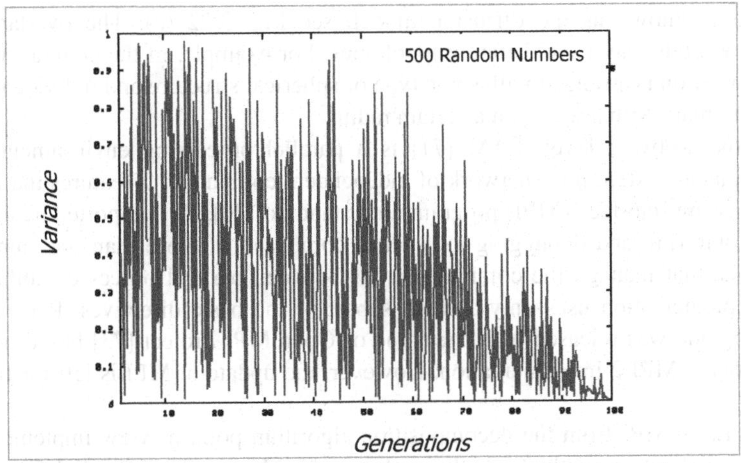

Fig. 3.21 Decreasing function for calculation of σ'

3.3.7 A Distributed Algorithm for Time Consuming Fitness Functions

Following the trends in the distributed computing domain we developed a distributed implementation of GENOM kernel adapting the original sequential enhanced GA using a standard message passing protocol, LAM/MPI [70],[71]. Once evolutionary algorithms consider populations of solutions, they are easily parallelizable [36],[26].

One of the most straightforward approaches is to have one global population with multiple processor units for evaluating individual solutions, see Fig. 3.22 (a). This scenario can be very useful for applications with heavy evaluation functions. Another method often used, known as the island model (Fig. 3.22 (b)), divides the global population in several subpopulations, each one executing its own evolutionary algorithm. Once in a while, one individual from one subpopulation receives permission to migrate to a neighbor subpopulation. Another approach allows the migration of data to a group of neighbors that share areas of interest. This

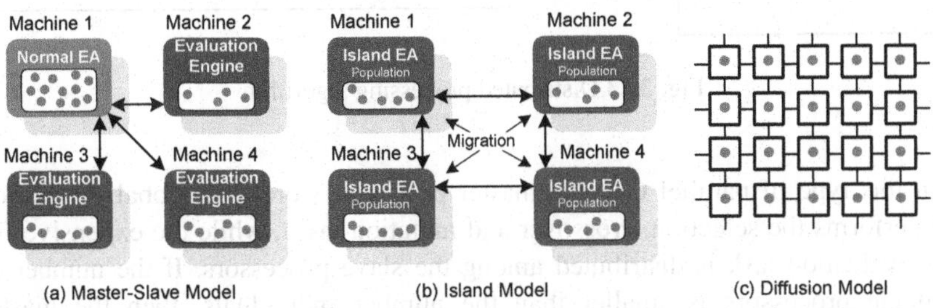

Fig. 3.22 Parallel architectures for EAs

method is known as the diffusion model, see Fig. 3.22 (c). The overlapping neighborhoods may have different topologies. For example, in the cellular model the population is arranged with some type of spherical structure and individuals allowed to mate with, are within a certain radius.

At the software level, LAM [71] is a parallel processing environment and development system for a network of independent computers. It features the Message-Passing Interface (MPI) programming standard [70],[72] supported by extensive monitoring and debugging tools. It is composed of more than one hundred functions that manage the communications between parallel processes, although this implementation uses only a small subset of the basic directives. Recently a new upgrade was released with the name of Open MPI Project [73] based on the open source MPI-2 implementation. However, the update of MPI is left for future work.

The LAM/MPI from the decomposition algorithm point of view implements a new parallelization method which combines the Master-Slave Parallel GA with the Coarse-Grain GA methods [74] in a network of independent computers or in a single processor with several cores, as illustrated in Fig. 3.23.

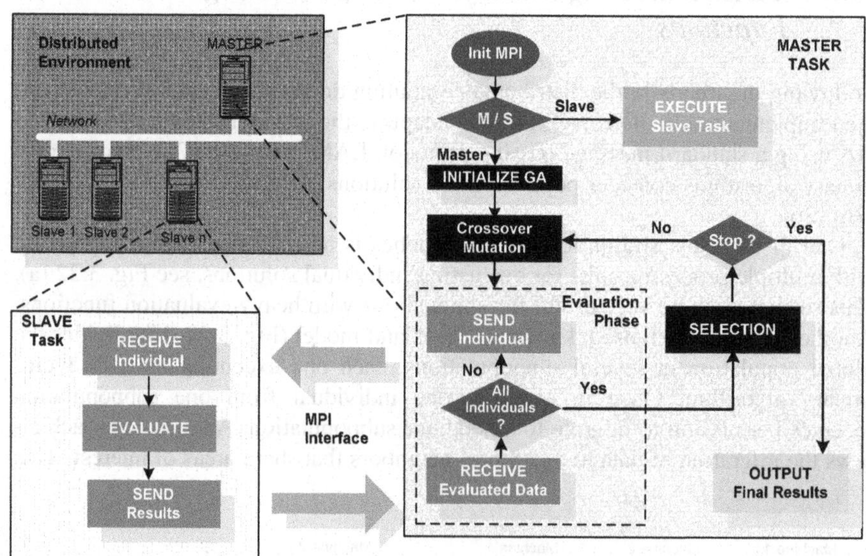

Fig. 3.23 Distributed processing algorithm

In this type of parallel GA, the master processor stores the global population and performs the selection, crossover and mutation tasks, while the expensive fitness evaluation task is distributed among the slave processors. If the number of available processors is smaller than the number individuals, then the master transfers another evaluation task to the free slaves, as soon as, they finish their tasks. The expected speed-up of the proposed parallel method, for expensive fitness evaluations, can increase nearly linear along with the number of used slave processors [26],[69],[75]-[76], as depicted in Fig. 3.24. The main advantage of this model resides in increasing the algorithm speed without introducing extra complexity. The results are in conformance with the expected theoretical model given in [26]. The real values are slightly below the linear speed-up reference line due to communications costs.

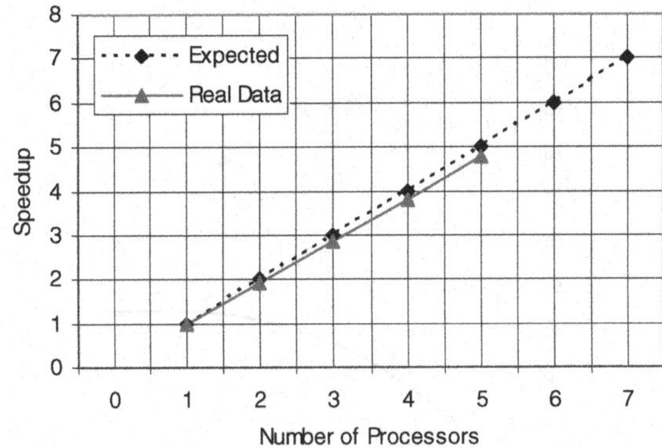

Fig. 3.24 Expected distributed processing speed-up tested with GENOM

The search space decomposition, described in Sect. 3.3.2, is an additional technique which can raise the algorithm performance when used together with parallelization. Basically, the master processor after decomposing the search space in small p^n problem subspaces, assigns to each slave processor the execution of one subspace optimization task and at the end the Master processor executes a final global optimization task having those chromosomes as the initial population.

The program code was relatively simple to adapt and it can easily keep the load balance. The inter-processor data communication overheads produced in this model is much less when compared with others. Besides, the method does not change the GA search behavior, so the conclusions for the serial GA Kernel can still be applied.

According to Cantu-Paz and Goldberg [26] the total execution time per generation of a parallel GA can be computed as:

$$Tp = \frac{Npop.Tf}{P} + \rho(P-1)Tc \tag{3.12}$$

where,

Tp = the the total execution time per generation
Npop = the number of chromosomes in the population
Tf = the time to evaluate the fitness of one chromosome
Tc = the average time to communicate with one processor

P = number of processors

ρ = parameter dependent on selection and parallelization method

It can be seen that the total execution time is composed of two terms, the first, refers to the time required to evaluate the fitness of the chromosomes and, the second, involves the total communication time. The speedup for a given number of processors can be computed by expression (3.13), where *T1* is the time for a single processor [36]:

$$Speedup = \frac{T1}{T_p} \qquad where,\ T_1 = \frac{Npop.Tf}{P} \qquad (3.13)$$

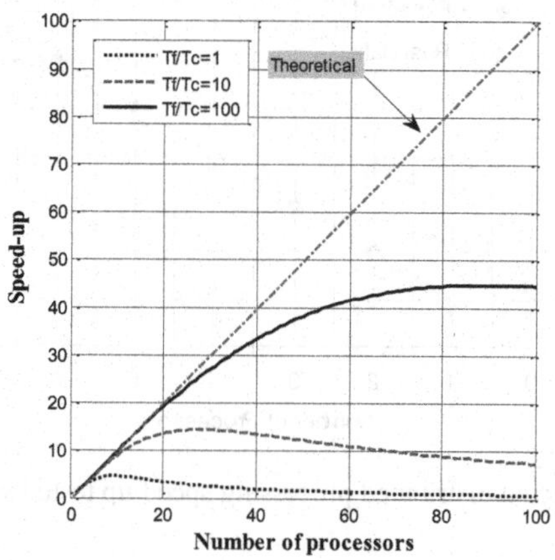

Fig. 3.25 Theoretical speedup of GAs with Npop = 100

That speedup depends on the ratio of the time to compute the fitness relative to the communication time (Tf /Tc), the number of processors, the population size, and the variable ρ, which depends on the details of the code and the parallelization technique. Here we use ρ = 1, which is appropriate for a master–slave GA application.

3.3.8 GENOM GA Attributes

Table 3.9 provides a description of the most commonly used techniques applied in EAs in terms of decisions that must be taken into account during an implementation of a particular EA and shows how GENOM fits in this domain. It is an extended variation of the table proposed in [27].

The attribute values represent some functionality that a designer wishes to integrate in the optimization framework. From the designer point of view, the set of attributes represents the decisions that will characterize a given system.

Table 3.9 Overview of the common used attributes in GENOM and EAs

Nº	Attribute			Attribute Values 1	2	3	4	5
1	**Solution Representation**	Encoding		Binary	Real-value	Integer-valued	Trees (GP)	Symbolic
		Length		Fixed	Variable			
2	**Fitness Function**			Single Objective	Single Objective, weight minimization	Multi-objective	Multi-objective Multi-Constraint	
2	**Population Initialization**	Mechanism		Random Generation	Grid initialization	Knowledge-based	LHS*	DOE*
		Population Size		Single and Fixed	Single and Variable	Multiple and Fixed	Multiple and Variable	
3	**Parent Selection**			Uniform	Ranking	Fitness Proportional	Tournament	Tournament with Feasibility-based
4	**Evolutionary Operators**	Mutation	Type	Uniform	Non-uniform	Gaussian		User defined: PCP*
			Rate	Fixed	Dynamic	Adaptive	Self-Adaptive	User defined: UNDO*
		Crossover	Type	Uniform	Single Crossover	Multiple Crossover	Parent Centric	User defined: CSM*
			Rate	Fixed	Random	Dynamic	Adaptive	
5	**Survival Selection Mechanism**			Truncation (μ, λ)	Truncation ($\mu + \lambda$)			
6	**Constraint Handling Method**			Penalty Function	Special representations and operators	Repair algorithms	Separation of objectives& constraints	Hybrid methods
7	**Population Diversity**			Penalty Function	Special Representations and Operators	Sharing	Crowding	Fuzzy Logic
8	**Hybridization Approach**			GA with Stochastic	GA with Neural Nets	GA with SVM	GA with Response Surface Models	GA with Krigging Model
9	**Parameter Control**			No Control	Deterministic Step Size Control	Adaptive Step Size Control	Self-Adaptive	
10	**Convergence Speed-up**			Distributed Processing Master-Slave archit.	Parallel with Migration	Diffusion Model		

(*) LHS – Latin Hyper Sampling ;
DOE – Design of Experiments;
PCP – Premature Convergence Prevention
CSM – Crossover with Standard Mutation
UNDO – Undo Function

Legend:
GENOM attribute (shaded)
EA General attribute (unshaded)

3.3.9 GENOM Optimization Methodology

One of the major challenges facing semiconductor companies today is how to increase yield. There are many factors that affect yield. Here, this focus is given mainly to those related to process, temperature and supply voltage variations inherent to any IC design environment and fabrication process. A corner point, as seen by an IC designer, is a combination of a particular technological model describing some process variations with a particular operating context representing some combination of operating conditions. The corner analysis represents the circuit performance analysis for all the combinations of extreme corner points as illustrated in Fig. 3.26.

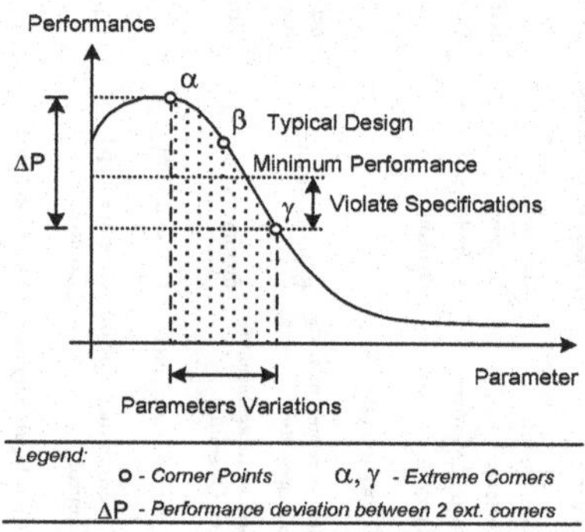

Fig. 3.26 Performance specification violation arouse from operational circuit deviations

Corners simulation is perhaps the most widely and less time-consuming method used to test process, temperature and voltage variations. Usually, a designer determines the worst case corners, or conditions, under which the circuit or system is expected to function and these are the minimum requirements to produce a robust design. This important requirement was taken into consideration in the development of GENOM optimization methodology.

The proposed methodology, illustrated in Fig. 3.27, is based on an enhanced GA kernel implemented in two optimization moments, a coarse and fine optimization. The first one executes a nominal optimization of analog building blocks and, in the second one the optimization is extended to deal with operational and process variations. This type of analysis is a mandatory task in any modern analog design process in order to get realistic solutions or robust designs.

3.3.9.1 Optimization Setup

The IC designer inserts the necessary input data in a database system assisted with the AIDA/GENOM design front-end [77] (see also Sect. 5.2). First, the circuit topology and technology files are selected from the IC database. Then, a specification page, dynamically generated based on the number and class of optimization variables and technological constraints, allows the setup of parameter bounds, etc.

Next, the required design goals, the circuit design constraints and the type of optimization procedure must be appropriately specified. In case of a circuit design optimization with corner analysis, the information like process and operation variations to be used, and their associated weight, should also be described. Finally, the job is executed in a single machine or in a distributed environment using the built-in distributed processing capability described in [69] and sub-Sect. 5.2.4.

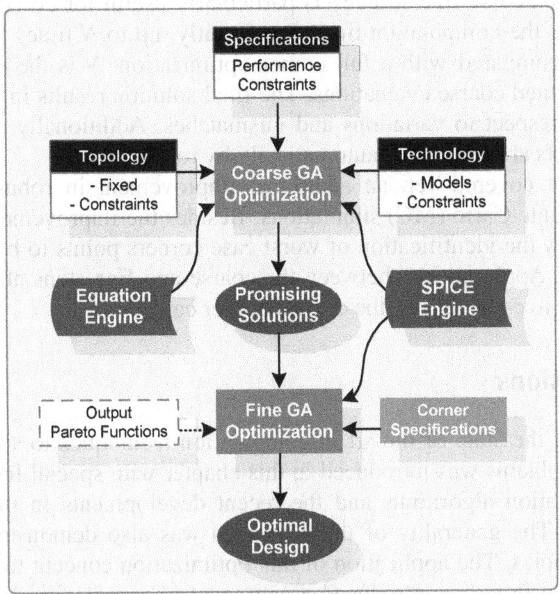

Fig. 3.27 Circuit/System optimization methodology

3.3.9.2 Coarse Optimization

The coarse optimization step executes a nominal optimization of analog building blocks considering only typical device models and typical working conditions. The purpose of the coarse optimization is to generate, in a fast way, a set of potential good solutions that will represent the initial population for the fine-tuning optimization process. Here, the optimization parameters can be relaxed, for example, one can use a multiple of the fine grid value, thus reducing the search space and increasing the probability of finding space regions with potentially valid solutions for the second optimization run.

3.3.9.3 Fine-Tuning Optimization

Here a fine grid optimization is performed taking as the initial population the ELITE chromosomes from the preceding step and having into account the given corners analysis specifications. This second optimization step is executed using the same optimization algorithm and using an expanded version of the algorithm fitness function defined in sub-Sect. 3.3.1. Now, the algorithm merit function assigns to each corner a weight proportional to its importance and fitness value.

The idea beyond this strategy is, on one hand, a common belief according to which one solution that verifies all corners must satisfy necessarily all performance specifications, including the typical case specs and, on the other hand, the influence of each corner condition will modify the performance measures in the boarder around the corner point reached by optimization in typical conditions.

This two-step coarse-fine strategy is particularly useful for corner analysis because it reduces the computation time significantly, up to N times the number of corners, when compared with a full corner optimization. N is the defined as the number of executed coarse evaluations. The final solution results in a more robust approach with respect to variations and mismatches. Additionally, the undesired sensitivity effects are attenuated automatically by robust design.

Although not covered yet, an additional improvement in robustness can be achieved by Monte Carlo (MC) simulations. Besides the improvement in terms of yield, they allow the identification of worst case corners points to be used later in the sizing loop. Applying MC between the coarse and fine steps allows the optimization kernel to consider just the critical corner cases.

3.4 Conclusions

An overview of the state of the art in computation techniques to solve nonlinear optimization problems was introduced in this chapter with special focus on evolutionary optimization algorithms and the recent developments in this field were also described. The generality of this approach was also demonstrated in some numerical examples. The application of this optimization concept to automate the analog IC design flow has introduced a new level of complexity. In this class of optimization algorithms, designers can control the setup parameters of the optimization problem and even some parameters from the optimization algorithm but do not have the keys to guarantee that the computed solution is really the optimal one. To ensure an efficient resolution of the optimization algorithm, designers have to formulate an adequate cost function and define efficient criteria to be used by the genetic operators. A new optimization tool called GENOM, based on a GA optimization kernel, has been designed to capture the design performance targets of an analog multi-objective multi-constrained IC design problem. The system was designed to incorporate the effect of process variations, this way the optimized circuit becomes tolerant to process variations increasing the yield.

References

[1] Medeiro, F., et al.: A Statistical optimization-based approach for automated sizing of analog cells. In: Proc. ACM/IEEE Int. Conf. Computer-Aided Design, pp. 594–597 (1994)

[2] Nye, W., Riley, D.C., Sangiovanni-Vincentelli, A., Tits, A.L.: DELIGHT.SPICE: An optimization-based system for the design of integrated circuits. IEEE Trans. Computer-Aided Design 7(4), 501–519 (1998)

[3] Michalewicz, Z.: Evolutionary computation techniques for nonlinear programming problems. International Transactions in Operational Research 1(2), 223–240 (1994)
[4] Zitzler, E.: Evolutionary algorithms for multi-objective optimization: Methods and applications. Ph.D. Thesis, Swiss Federal Institute of Technology (ETH), Zurich (1999)
[5] Fonseca, C.M., Fleming, P.J.: An overview of evolutionary algorithms for multi-objective optimization. In: Proc. Congress on Evolutionary Computation, vol. 3(1), pp. 1–16 (1998)
[6] NEOS Guide, Optimization Technology Center, Department of Energy, Northwestern University (2005), `http://www.ece.northwestern.edu/OTC` (Accessed March 2009)
[7] Dantzig, G.B.: Linear programming and extensions. Princeton University Press, Princeton (1963)
[8] Beasley, J.E.: Advances in linear and integer programming. Oxford Science Publications, Oxford (1996)
[9] Bertsekas, D.P.: Nonlinear programming, 2nd edn. Athena Scientific,Belmont (1998)
[10] Constraint programming, Artificial intelligence applications institute. The University of Edinburgh (2007), `http://www.aiai.ed.ac.uk/` (Accessed March 2009)
[11] Deb, K.: An efficient constraint handling method for genetic algorithms. Computer Methods in Applied Mechanics and Engineering 186, 311–338 (2000)
[12] Coello, C.A.C.: Theoretical and numerical constraint handling techniques used with evolutionary algorithms: A survey of the state of the art. Computer Methods in Applied Mechanics and Engineering 191, 1245–1287 (2002)
[13] Mezura-Montes, E., Velázquez-Reyes, J., Coello, C.A.C.: Promising infeasibility and multiple offspring incorporated to differential evolution for constrained optimization. In: Proc. GECCO, pp. 225–232 (2005)
[14] Mahfoud, S.W., Goldberg, D.E.: Parallel recombinative simulated annealing: A genetic algorithm. Parallel Computing 21, 1–28 (1995)
[15] Metropolis, N., Rosenbluth, A.W., Rosenbluth, M.N., Teller, A.H., Teller, E.: Equation of state calculations by fast computing machines. Journal of Chemical Physics 21(6), 1087–1092 (1953)
[16] Kirkpatrick, S., Gerlatt, C.D., Vecchi, M.P.: Optimization by simulated annealing. Science (1983), doi: 10.1126/science.220.4598.671
[17] Schaffer, J.D.: Some experiments in machine learning using vector evaluated genetic algorithms. Ph.D. dissertation, Vanderbilt University, Nashville, TN (1984)
[18] Fonseca, C.M., Fleming, P.J.: Genetic algorithms for multi-objective optimization: Formulation. Discussion and Generalization. In: Proc. 5th International Conference on Genetic Algorithms, pp. 141–153 (1993)
[19] Fonseca, C.M., Fleming, P.J.: Multi-objective optimization and multiple constraints handling with evolutionary algorithms–Part II: Application example. IEEE Trans. Systems, Man, and Cybernetics: Part A: Systems and Humans, 38–47 (1998)
[20] Fonseca, C.: Multiobjective genetic algorithm with application to control engineering problems, Ph.D. Thesis, The University of Sheffield (1995)
[21] Deb, K., Pratap, A., Agrawal, S., Meyarivan, T.: A fast and elitist multi-objective genetic algorithm: NSGA-II. IEEE Trans. Evolutionary Computation 6, 182–197 (2002)
[22] Horn, J., Nafploitis, N., Goldberg, D.E.: A niched pareto genetic algorithm for multi-objective optimization. In: Proc. 1st IEEE Conference on Evolutionary Computation, pp. 82–87 (1994)

[23] Wolpert, D.H., Macready, W.G.: No free lunch theorems for optimization. IEEE Trans. Evolutionary Computation 1, 67–82 (1997)

[24] Goldberg, D.E.: Genetic algorithms in search, optimization and machine learning. Addison-Wesley, Reading (1989)

[25] Michalewicz, Z.: Genetic algorithms + data structure = evolution programs, 3rd edn. Springer, Berlin (1996)

[26] Cantu-Paz, E., Goldberg, D.E.: On the scalability of parallel genetic algorithms. IEEE Trans. Evolutionary Computation 7, 429–449 (1999)

[27] Kicinger, R., Arciszewski, T., De Jong, K.A.: Evolutionary computation and structural design: A survey of the state of the art. Computers & Structures 83(23), 1943–1978 (2005)

[28] Zilouchian, A., Jamshidi, M.: Intelligent control systems using soft computing methodologies. CRC Press LLC (2001)

[29] Liang, J., McConaghy, T., Kochlan, A., Pham, T., Hertz, G.: Intelligent systems for analog circuit design automation: A survey (2001), `http://archived.techonline.com/` (Accessed March 2009)

[30] Holland, J.H.: Adaptation in Natural and Artificial Systems. The University of Michigan Press, Ann Arbor (1975)

[31] Back, T., Hoffmcister, F., Schwefel, H.P.: A survey of evolution strategies. In: Proc. 4th Int. Conf. on Genetic Algorithms, pp. 2–9 (1991)

[32] Fogel, L.J., Owens, A.J., Walsh, M.J.: Artificial intelligence through simulated evolution. Wiley, Chichester (1966)

[33] Fogel, D.B.: Evolutionary computation: Towards a new philosophy of machine intelligence. IEEE Press, New York (2000)

[34] Koza, J.R.: Genetic Programming: On the Programming of Computers by Means of Natural Selection. MIT Press, Cambridge (1992)

[35] Koza, J.R., Keane, M.A., Streeter, M.J., Mydlowec, W., Yu, J., Lanza, G.: Genetic programming IV: Routine human-competitive machine intelligence. Kluwer Academic Publishers, Dordrecht (2003)

[36] Haupt, R.L., Haupt, S.E.: Practical genetic algorithms. Wiley, New York (1998)

[37] Dorigo, M., Maniezzo, V., Colorni, A.: The ant system: Optimization by a colony of cooperating agents. IEEE Trans. Systems, Man, and Cybernetics-Part B 26(1), 29–41 (1996)

[38] Kennedy, J., Eberhart, R.: Particle swarm optimization. In: Proc. IEEE Int. Conf. on Neural Networks, pp. 1942–1948 (1995)

[39] Ocenasek, J.: Parallel estimation of distribution algorithms. Ph.D. dissertation, Faculty of Information Technology, Brno University of Technology (2002)

[40] Larrañaga, P., Lozano, J.A.: Optimization by learning and simulation of probabilistic graphical models. In: Parallel Problem Solving from Nature, PPSN VII (2002), `http://www.sc.ehu.es/ccwbayes/` (Accessed March 2009)

[41] Larrañaga, P., Lozano, J.A.: Estimation of distribution algorithms: A new tool for evolutionary computation. Kluwer Academic Publishers, Norwell (2001)

[42] Price, K., Storn, R.: Differential evolution - a simple and efficient heuristic strategy for global optimization over continuous spaces. Journal of Global Optimization 11, 341–359 (1997)

[43] Reynolds, R.G.: An introduction to cultural algorithms. In: Sebald, A.V., Fogel, L.J. (eds.) Proc. 3rd Annual Conference on Evolutionary Programming, pp. 131–139 (1994)

[44] Moscato, P.: On evolution, search, optimization, genetic algorithms and martial arts: Towards memetic algorithms. Caltech Concurrent Computation Program, C3P Report 826 (1989)

[45] Krasnogor, N., Smith, J.E.: A tutorial for competent memetic algorithms: Model, taxonomy and design issues. IEEE Trans. Evolutionary Computation 9(5), 474–488 (2005)

[46] Ong, Y.S., Krasnogor, N., Ishibuchi, H.: Special Issue on Memetic Algorithms. IEEE Trans. Systems, Man and Cybernetics - Part B 37(1) (2007)

[47] Ong, Y.S., Nair, P.B., Keane, A.J., Wong, K.W.: Surrogate-assisted evolutionary optimization frameworks for high-fidelity engineering design problems. In: Knowledge Incorporation in Evolutionary Computation, pp. 307–332. Springer, Heidelberg (2004)

[48] Bosman, P.A., Thierens, D.: Exploiting gradient information in continuous iterated density estimation evolutionary algorithms. Tech. Rep. UU-CS-2001-53, Universiteit Utrecht (2001)

[49] Mezura-Montes, E., Coello, C.A.C.: Adding a diversity mechanism to a simple evolution strategy to solve constrained optimization problems. In: Proc. Congress on Evolutionary Computation, vol. 1, pp. 6–13 (2003)

[50] Streichert, F., Stein, G., Ulmer, H., Zell, A.: A Clustering based niching EA for multimodal search spaces. In: Proc. 6th International Conference Evolution Artificielle, pp. 169–180 (2003)

[51] Kim, J.K., Cho, D.H., Jung, H.K., Lee, C.G.: Niching genetic algorithm adopting restricted competition selection combined with pattern search method. IEEE Trans. Magnetics 38(2), 1001–1004 (2002)

[52] Knowles, J., Corne, D.: The pareto archived evolution strategy: A new baseline algorithm for multi-objective optimization. In: Proc. Congress on Evolutionary Computation, pp. 98–105. IEEE Service Center, New Jersey (1999)

[53] Ishibuchi, H., Nojima, Y., Doi, T.: Comparison between single-objective and multi-objective genetic algorithms: Performance comparison and performance measures. In: Proc. Congress on Evolutionary Computation, pp. 1143–1150 (2006)

[54] Mezura-Montes, E., Coello, C.A.: Multiobjective-Based Concepts to Handle Constraints in Evolutionary Algorithms. In: Proc. 4th Mexican International Conference on Computer Science, pp. 192–199 (2003)

[55] Jin, Y.: A comprehensive survey of fitness approximation in evolutionary computation. Soft Computing Journal 9(1), 3–12 (2005)

[56] Bäck, T., Fogel, D., Michalewicz, Z.: Handbook of evolutionary computation. Oxford University Press, Oxford (1997)

[57] Eiben, A.E., Hinterding, R., Michalewicz, Z.: Parameter control in evolutionary algorithms. IEEE Trans. Evolutionary Computation 3(2), 124–141 (1999)

[58] Gabriel, E., Fagg, G., Bosilca, G., Angskun, T., Dongarra, J., et al.: open MPI: Goals, concept, and design of a next generation MPI implementation. In: Proc. 11th European PVM/MPI Users' Group Meeting, Hungary (2004)

[59] Andersson, J.: A survey of multiobjective optimization in engineering design. Tech. Rep. No. LiTH-IKP-R-1097, Dept. of Mechanical Engineering, Linköping University (2000)

[60] Andersson, J.: Multiobjective optimization in engineering design applications to fluid power systems, Ph.D. Thesis no 675, Linköping University, Linköping (2001)

[61] Messac, A., Sundararaj, G.J., Tappeta, R.V., Renaud, J.E.: The ability of objective functions to generate non-convex pareto frontiers. American Institute of Aeronautics and Astronautics Journal 38(3), 1084–1091 (2000)

[62] Messac, A.: Physical programming: Effective optimization for computational design. American Institute of Aeronautics and Astronautics Journal 34(1), 149–158 (1996)

[63] Messac, A., Wilson, B.: Physical programming for computational control. American Institute of Aeronautics and Astronautics Journal 36(2), 219–226 (1998)

[64] Michalewicz, Z., Michalewicz, M.: Evolutionary computation techniques and their applications. In: Proc. IEEE International Conference on Intelligent Processing Systems, vol. 1, pp. 14–25 (1997)

[65] Czarn, A., MacNish, C., Vijayan, K., Turlach, B., Gupta, R.: Statistical exploratory analysis of genetic algorithms. IEEE Trans. Evolutionary Computation 8(4), 405–421 (2004)

[66] McKay, M.D., Conover, W.J., Beckman, R.J.: A comparison of three methods for selecting values of input variables in the analysis of output from a computer code. Technometrics 21, 239–245 (1979)

[67] Antony, J., Somasundarum, V., Fergusson, C.: Applications of taguchi approach to statistical design of experiments in Czech Republican industries. International Journal of Productivity and Performance Management 53(5), 447–457 (2004)

[68] Trygg, J., Wold, S.: Introduction to statistical experimental design. Editorial (2002), `http://www.acc.umu.se/~tnkjtg/Chemometrics/Editorial` (Accessed March 2009)

[69] Barros, M., Neves, G., Guilherme, J., Horta, N.C.: A distributed enhanced genetic algorithm kernel applied to a circuit/level optimization E-Design environment. In: Proc. Design of Circuits and Integrated Systems, pp. 20–24 (2004)

[70] MPI, The message passing interface (MPI) Standard (1995), `http://www-unix.mcs.anl.gov/mpi/index.htm` (Accessed March 2009)

[71] LAM/MPI, LAM/MPI parallel computing (2007), `http://www.lam-mpi.org` (Accessed March 2009)

[72] MPI Primer/Developing with LAM, Ohio Supercomputer Center, `http://parallel.ksu.ru/ftp/mpi/LAM/lam61.doc.pdf.gz` (Accessed March 2009)

[73] Open MPI, Open MPI: Open source high performance computing (2007), `http://www.open-mpi.org` (Accessed March 2009)

[74] Zhang, L., Kleine, U.: A novel analog layout synthesis tool. In: Proc. IEEE Int. Symposium on Circuits and Systems, vol. 5, pp. 101–104 (2004)

[75] Chang, C., Lin, C.: LIBSVM: A library for support vector machines (2001), `http://www.csie.ntu.edu.tw/~cjlin/libsvm` (Accessed March 2009)

[76] Martin, K., Johns, D.: Analog integrated circuit design. John Wiley & Sons Inc., Chichester (1996)

[77] Barros, M., Neves, G., Horta, N.C.: AIDA: Analog IC design automation based on a fully configurable design hierarchy and flow. In: Proc. 13th IEEE International Conf. on Electronics, Circuits and Systems, pp. 490–493 (2006)

4 Enhanced Techniques for Analog Circuits Design Using SVM Models

Abstract. In order to improve the relatively slow convergence of GA, in the presence of large search spaces, and reduce the high consuming time of evaluation functions in analog circuit design applications, this chapter will discuss the use of learning algorithms. These algorithms explore the successive generation of solutions, learn the tendency of the best optimization variables and will use this knowledge to predict future values. In other words, these techniques employ data mining theory, used to manage large databases and huge amount of internet information, to discover complex relationships among various factors and extract meaningful knowledge to improve the efficiency and quality of decision making. In this chapter a new hybrid optimization algorithm is presented together with a design methodology, which increases the efficiency on the analog circuit design cycle. This new algorithm combines an enhanced GA kernel with an automatic learning machine based on SVM model (GA-SVM) which efficiently guides the selection operator of the GA algorithm avoiding time-consuming SPICE evaluations of non-promising solutions. The SVM model is here defined as a classification model used to predict the *feasibility region* in the presence of large, non-linear and constraints search spaces that characterize analog design problems. The SVM modeling attempts to constraint the search space in order to accelerate the search towards the feasible region ensuring a proper operation of the circuit.

4.1 Learning Algorithms Overview

Data mining consists of exploring data in order to discover unknown patterns and meaningful relationships in data, which may be used to make valid predictions. Within this technology data play an important role and the knowledge, extracted by the use of pattern recognition technologies as well statistical and mathematical techniques are the driven force in the new decision support systems. The adoption of this technology can increase the productivity in business or in the process where it was applied, since the same goals could be achieved or even improved with less investment in efforts and resources.

The technology behind data mining techniques is mostly based on inductive learning [1], where a model is constructed by generalizing from an adequate number of training samples collected from an historical database or coming from an experiment in which the sample is tested. Once built, the trained model can be applied to unseen examples to predict future trends and behaviors. This typical learning scenario is illustrated in Fig. 4.1 and it is known as a supervised learning approach. This differs from other approaches in what concerns to the feedback,

M.F.M. Barros et al.: Analog Circuits and Systems Optimization, SCI 294, pp. 89–107.
springerlink.com

received from its process during the learning stage. For example, in the reinforcement learning the feedback signal does not contain the knowledge of the environment, which is supplied to the learning machine in the supervised learning. Instead, the learning machine only receives a rating of its performance, often called reinforcement signal. In the unsupervised learning approach, the learning machine does not receive any feedback information at all, only the input samples. The learning machine is charged to reveal properties or knowledge hidden in the data, e.g. associating these data into groups or classes based on correlation of samples.

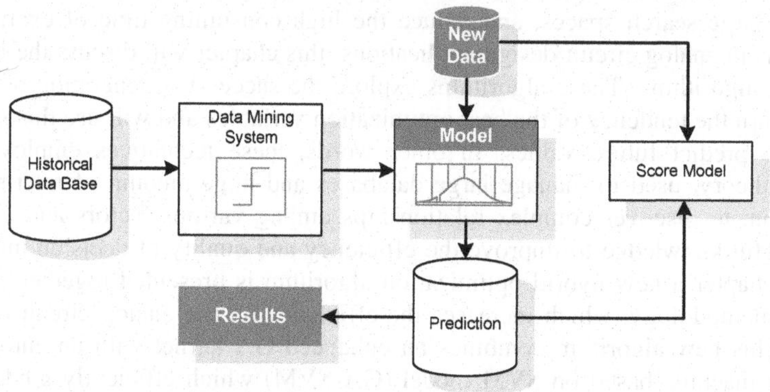

Fig. 4.1 Supervised learning approach

The data mining tools were originally developed to answer to specific problems in several areas of application and different knowledge domains, and so they inherited special characteristics that make each specific technique tailored to some type of problem. The most usual of these are:

- Classification and regression. These classes embrace the largest number of problems in the data mining domain [2]. In the classification problems, the learning machine creates a model to predict the class membership to which an entity belongs to, whereas in the regression case the model aims to predict a real-value variable based on the relationship between the other variables, assuming a linear or nonlinear relationship.
- Association and sequencing. Also known as market basket analysis, these techniques create models to discover hidden patterns of behavior, correlations among a set of objects generating an output in the form of descriptive rules, e.g., "75% of the customers who buy milk also buy bread and eggs". The sequencing technique is very similar to an association technique, but it includes description rules with information of time in the final analysis.
- Clustering. This technique seeks to identify a set of groups or clusters that defines the given data. Basically, it groups together entities or data points with similar behavior or properties, and creates different groups for dissimilar entities.

Advances in data mining were boosted by the progress in the fields of artificial intelligence (AI) and statistics. Fig. 4.2 provides a description of some of the most common data mining approaches used nowadays. Below, these techniques are briefly described.

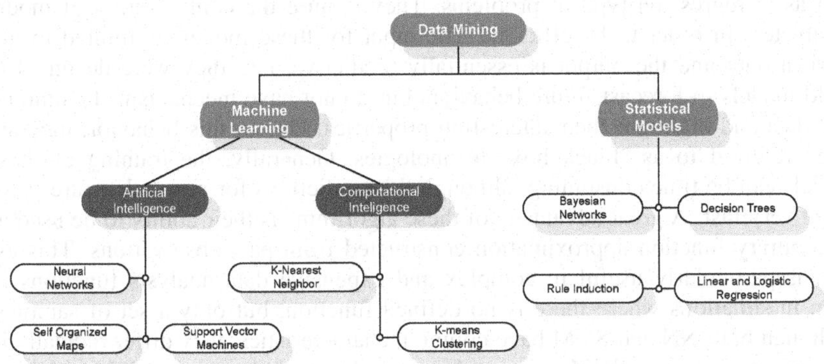

Fig. 4.2 Data mining technology

The regression technique implements a model based on observed data to forecast the output effect of a data item on the modeled system. In the simplest case, regression uses standard statistical techniques such as linear regression, which is modeled by a strait line that best fits the data and lately uses this line to predict values. The optimum model is obtained through the line that minimizes the sum of the square error from each data sample. The linear regression equation is described in Fig. 4.3.

However, for many real-world problems, predictions are very difficult to obtain because they may depend on complex interactions between multiple predictor

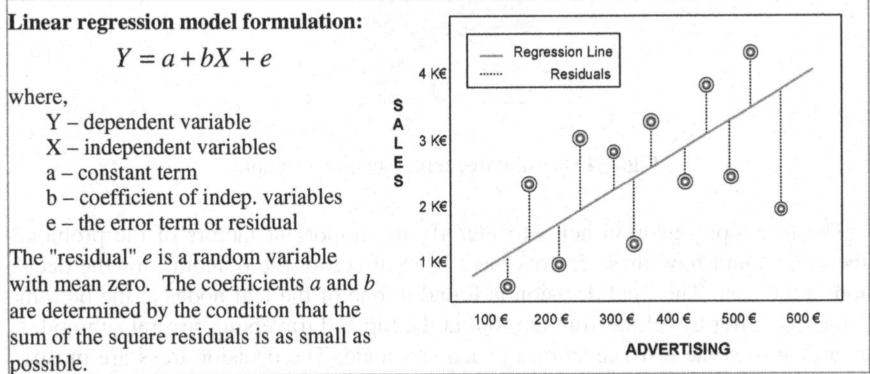

Fig. 4.3 Linear regression

variables. Therefore, more sophisticated algorithms are used for these cases such as, logistic regression, decision trees, neural nets and support vector machines.

Neural networks (NN) [3-8] are inspired on an early model of human brain function, whereas support vector machines (SVM) had their inspiration on the statistical learning theory [9]. Both have proven great efficiency either in classification as in regression type of problems. They require the configuration of model parameters in order to be efficient. The input for these models is limited to numerical data and the output is essentially predictive, i.e., they were designed to build models to forecast future behaviors but do not have mechanisms to summarize data and highlight their interesting properties. Due to this behavior, they are often referred to as "black box" technologies. Generally, the training of these models can be time consuming, although the predictions for new values are processed very fast. A great advantage of these algorithms is their ability to be used as an arbitrary function approximation constructed from past observations. This aspect is particularly useful in complex and expensive data analysis functions or even in situations where there is no defined function, but only a set of samples. Although both NN and SVM have common characteristics, they differ radically in one important aspect: SVM training always finds a global minimum [9].

The decision tree is a technique in which the resulting model is represented by graphic structure in a form of tree. One overview of this representation is illustrated in Fig. 4.4.

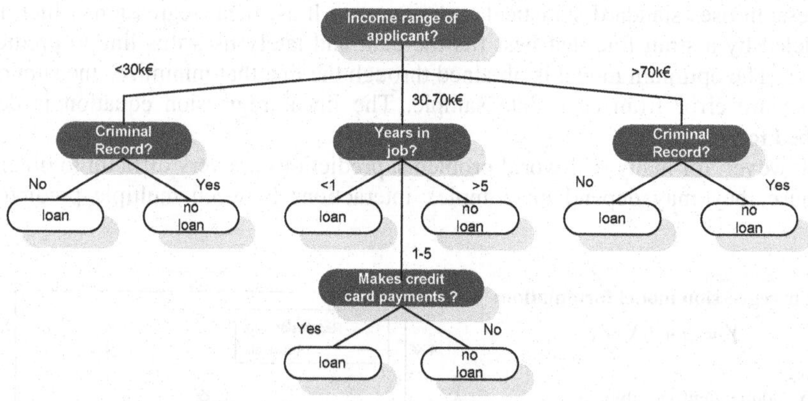

Fig. 4.4 Decision tree representation example

The tree representation helps to identify the important factors of the problem (the nodes) and how these factors have been affecting the outcomes of the decision in the past. The final decision is found in one of the leaf nodes at the bottom of the tree, after traveling from the root at the top and traversing several sub-nodes according to some test execution in each sub-node. The decision trees are mostly used for classification. The graphical representation is an attractive characteristic because it is easy to understand, which makes this technique become one of the most popular tools for data mining problems.

K-nearest neighbor (KNN) [10] is also a predictive technique suited for classification models. Unlike the other predictive techniques, it has no training phase once the training data represents simultaneously the model, thus models tend to be very large. The predictions for a new sample is done by looking for the group of similar characteristics and calculate the outcome value based on the most predominate class ("*k*" means the number of the nearest points with similar characteristics). The definition of this model is associated with a metric to measure the distances. The choice of metric is an important specification to take into account because the performance of the model depends on it.

K-means [10] is one of the simplest unsupervised learning algorithms tailored to solve the clustering problems. It is used to classify data, following a procedure that groups a given data set through a certain number of clusters (assume k clusters or subsets) defined by the user. The grouping routine minimizes the sum of squares of distances between data and the corresponding cluster centroid. When all samples have been assigned to a group which has the closest centroid, the algorithm recalculates the positions of the K centroids and repeats the process until the centroids get a stationary phase. Despite the simplicity, the k-means algorithm is also significantly sensitive to the initial randomly selected cluster centers and sometimes misses to find the most optimal configuration related to the global objective function minimization expressed in Fig. 4.5.

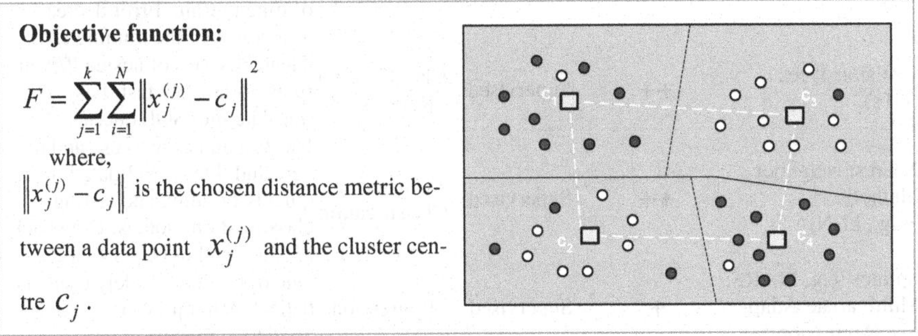

Objective function:
$F = \sum_{j=1}^{k}\sum_{i=1}^{N}\left\|x_i^{(j)} - c_j\right\|^2$
where,
$\left\|x_i^{(j)} - c_j\right\|$ is the chosen distance metric between a data point $x_i^{(j)}$ and the cluster centre c_j.

Fig. 4.5 K-means objective function

The next group of techniques has its origin in the *Naïve-Bayes* algorithm which uses the computation of probabilities as the main tool to make predictions. Naïve-Bayes is a classification technique that is not only predictive but also owns a descriptive feedback that describes the basic features and the interesting properties of the data. This approach assumes the statistically independence of all the independent variables which may not be true and is tailored to deal with categorical problems. The categorical limitation can be overcome to handle continuous data using bracket techniques that determine categories defined by limits of continuous data. Although technically simple to implement, the selection of the ranges can have a dramatic impact on the quality of the final model. The Naïve-Bayes concept is based on the relationship between dependent and independent variables and produces conditional probabilities derived from observed frequencies in the training data. Extending the Bayesian technique to capture the interactions between pairs of non-independent columns is also possible, although the complexity and storage capacity will increase a lot. However, in its simple form (assuming independence of variables) Naïve-Bayes is considered an easy and time efficient exploratory tool.

The Table 4.1 briefly summarizes some of the major characteristics of the learning algorithms presented in Fig. 4.2.

Table 4.1 Classification of data mining techniques

Methods	***Easy of use & understand***	***Class***	***Problems***	***Notes***
Support Vector Machines (SVMs)	-	Supervised	Classification Regression	SVMs are considered one of the most effective machine learning tool with the ability to represent non-linear relationships and produce models that generalize well to unseen data. SVM training always finds a global minimum.
Artificial Neural Nets (ANNs)	-	Supervised	Classification Regression Clustering	Relation between weights and variables is difficult to interpret. Dificult to build the network structure. Require large amounts of time to train. Error decreases as a power of the training size.
Decision Trees (DTs)	++	Supervised	Classification Regression	Clear. A series of nested if/then rules. Relatively fast. Easy to translate into SQL queries
Nearest Neighbor Methods (e.g., kNN)	++	Supervised	Clustering Classification	It is fast and easy to use and understand. Ideal candidate for quickly building and testing classification models. Drawback: Models tend to be very large.
Splines (e.g. MARS: Multivariate Adaptive Regression S.)	+	Supervised	Regression	One of the most widely used statistical techniques for creating predictive models
Logistic Regression	+	Supervised	Regression	One of the most widely used statistical techniques for creating predictive models.
Rule Learning	+	Supervised Unsupervised	Classification	Understandable. The computation of probabilities of all combinations can be expensive!
K-means clustering	++	Unsupervised	Clustering Classification	Simplicity. Sometimes misses the most optimal configuration and is sensitive to the initial cluster centers.
Self organized maps (SOM)	-	Unsupervised	Classification	Similar to feed-forward neural net except that there is one output for every hidden layer node.
Bayesian networks	+	Supervised	Classification	Limits their inputs to categorical data.

4.1.1 SVM Classification Overview

SVMs belong to a class of supervised learning algorithms which are able to acquire knowledge from previous experiences and to apply the knowledge to predict future values [1],[9], [11-12]. This process is known as *memorization* and *generalization*. The modeling presented here is based on a supervising SVM approach to the two-class classification problem, where a set of training data of the form $S=\{(x_i; y_i),\ldots,(x_n; y_n)\}$ is observed, and the input $x_i \in X \subset R^d$ is a d-dimensional feature vector and the output $y_i \in \{+1,-1\}$ is the class label of x_i. The main goal is to train a discriminate function, which will be used to predict the labels for new inputs, minimizing the probability of classification errors.

Generally, the support vector classifier is implemented in a two step process. First, it is applied the kernel "*trick*", which provides a nonlinear mapping of the vectors x_i into a higher dimensional *feature space*. In the second step, a decision boundary hyperplane is created based on the *maximal-margin principle*. This process is illustrated in Fig. 4.6 where the input space of two classes originally inseparable, is mapped into a feature space, making it possible the separation of the two classes in a linear way.

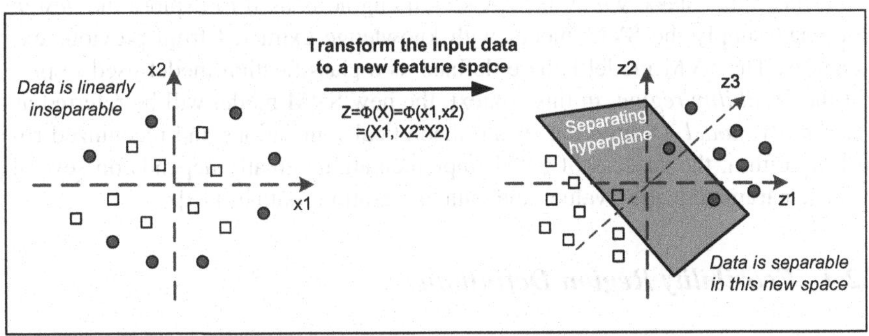

Fig. 4.6 Separating the data in a feature space

The SVM learning algorithm finds the *optimal separating hyperplane* (OSH) that maximizes the distance between the decision boundary between the two class groups and the closest point to the boundary, known as the margin, as illustrated in Fig. 4.7. The decision boundary points overlapping the margins are called *support vectors*. Support vectors are the most relevant in the decision process. The separating hyperplane in the feature space can correspond to a nonlinear decision boundary in the original input space. A more extended background on SVM concepts, issues and formulation, is presented in Appendix D.

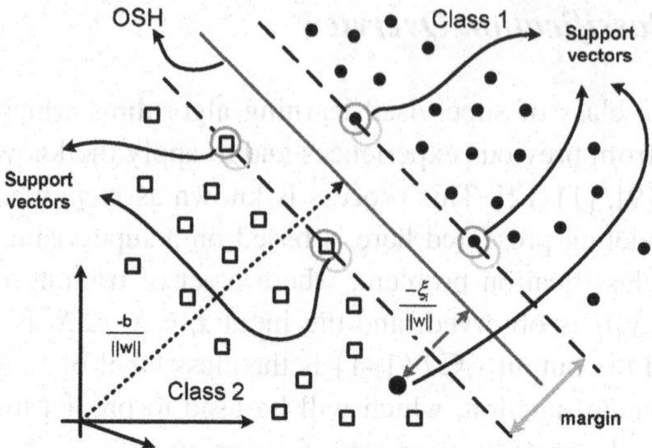

Fig. 4.7 Illustration of OSH hyperplane, margin and support vectors concept

4.2 GA-SVM Optimization Approach

In this book the supervised learning algorithm belonging to the class of machine learning algorithm called SVM, was adopted to work together with the selected GA approach. The GA-SVM methodology explores the properties of *the sizing rules method*, commonly used in analog circuits, and produces a *feasibility* model of the functional space while the GA search engine is used to explore the *design space* and supply the SVM model with knowledge extracted from previous experiences. The SVM model is here defined as a classification model used to predict the *feasibility region*, in this context, the new SVM model will be referred as a *feasibility model*. Despite the strong theoretical foundations and recognized robust algorithm, the success of SVM implementations greatly depends on several intrinsic parameterization values and data preparation routines [13].

4.2.1 Feasibility Region Definition

One problem often found with numerical optimization methods is the generation of results considered *pathological*, that is, a result that on the one hand meets all specifications but on the other hand fails some basic design requirements (e.g. saturation of certain transistors) [14], leading to a malfunction circuit. This inconvenient behavior is derived from insufficient design specifications, where a circuit optimization problem is considered as a black box with a number of design parameter constraints and performances constraints. Expert IC designers learn how to deal with *pathological* sizing by manually constrain the circuit to ensure proper biasing and good behavior of performance metrics. For example, fixing all transistor lengths and applying device matching conditions is a common practice employed in analog circuit design. The methodology which attempts to automatically constrain a circuit in order to ensure proper operation is called the *sizing rules method* [15]. Applying this methodology not only avoids the pathological designs but also improves the behavior of performance metrics and reduces sensitivity to operating conditions and process variations [16].

A generalized view of the sizing rules methodology points to the use of inequality constraints on electrical parameters (voltages and currents) in order to ensure the correct circuit operation. For example, [14],[17] introduces the concept of functional constraints and applies this concept to a simple CMOS current mirror. Functional constraints are a set of additional specifications defined analytically with a strong dependence on the application and the technology as illustrated in Fig. 4.8. This approach can be extended to other sub-circuits in order to determine a set of functional constraints. DELIGHT.SPICE [18] and the FRIDGE [19] tools were the first to take into account these concepts.

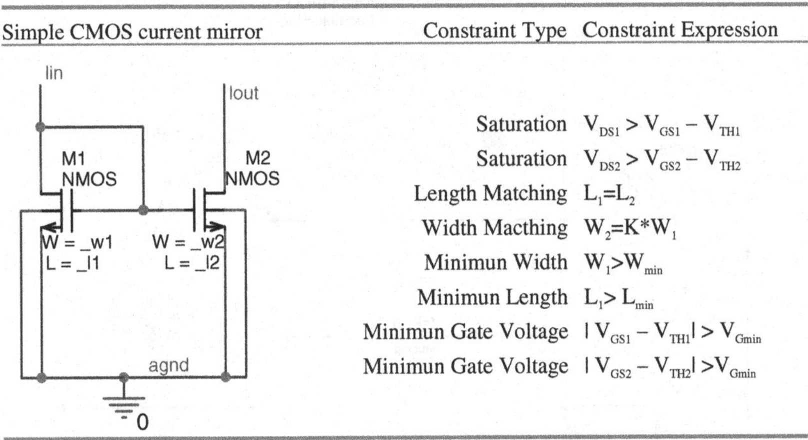

Simple CMOS current mirror
Iin
Iout
M1
NMOS
M2
NMOS
W = _w1
L = _l1
W = _w2
L = _l2
agnd
0
Constraint Type Constraint Expression
Saturation $V_{DS1} > V_{GS1} - V_{TH1}$
Saturation $V_{DS2} > V_{GS2} - V_{TH2}$
Length Matching $L_1=L_2$
Width Macthing $W_2=K*W_1$
Minimun Width $W_1>W_{min}$
Minimun Length $L_1> L_{min}$
Minimun Gate Voltage $| V_{GS1} - V_{TH1}| > V_{Gmin}$
Minimun Gate Voltage $| V_{GS2} - V_{TH2}| >V_{Gmin}$

Fig. 4.8 Functional constraints on a CMOS current mirror

In summary, the sizing rules methodology imposes some constraints not only in the *design space*, formed by the device sizes, but also in the *functional space*. In this context, the search space is decomposed in design space and functional space as illustrated in Fig. 4.9.

In a traditional optimization approach there is a mapping between a point in design space *d* (Fig. 4.9a) and a set of performances in performance space, *p* (Fig. 4.9c). In order to find a solution that satisfies the performance and functional constraints, usually, it requires the computation of many points from the design space. The achieved solution may result in a pathological case, in this condition, the result is not feasible (I in Fig. 4.9c). In the same way, the subspace of d defined by the interception of all functional constraints, the functional space f, may also produce *pathological* solutions, this time all functional constraints are satisfied but misses some performances specs (II in Fig. 4.9c). The *feasible region* defines a set of points in design space that satisfies both the performance constraints, as well as, the functional constraints. The multidimensional subspace of design parameters which fulfills all functional constraints is called in this work, the *feasibility space* (Fig. 4.9b) and the mapping of this space in performance space in called the *feasibility region* (Fig. 4.9c). If the multidimensional *feasible space* is known the computation time can be highly reduced.

Throughout this chapter, a new method which explores the properties of the sizing rules method and learning algorithms is developed in order to build a model for the functional feasibility space, the *feasibility model*. The aim of this approach is to accelerate the search towards the *performance feasible region* ensuring a proper operation of the analog circuits.

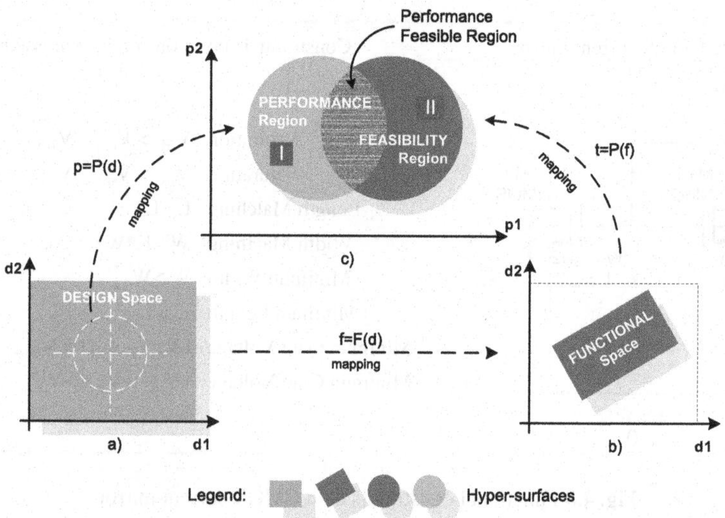

Fig. 4.9 Abstraction of analog circuit feasibility region

4.2.2 Methodology Overview

The evolutionary search algorithms in general have a common behavior. They cyclically generate new moves from the most fitness samples, evaluate them and then discard the less fitness ones. The less fitness offspring information is never used to decide what the next move should be or what path should be followed in getting to a local optimum. Rather than discarding information about the search, this new strategy uses all information from the evolutionary process to help us to make predictions about new data and improve the efficiency of the search algorithm. The new GA-SVM approach incorporates a learning model in the GA optimization cycle based on SVMs. The original GENOM optimization architecture is expanded with modeling capabilities as illustrated by Fig. 4.10.

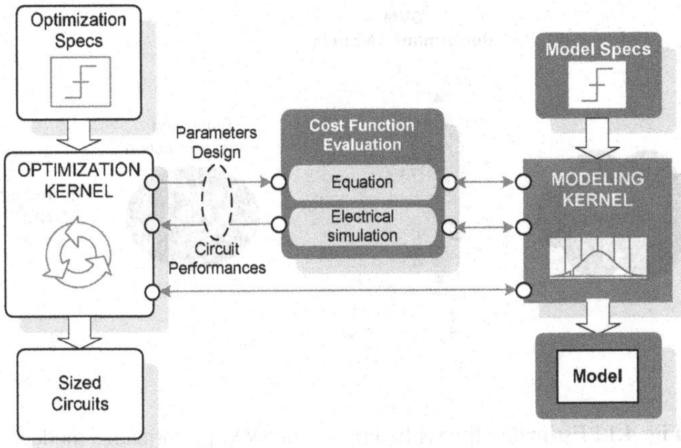

Fig. 4.10 Optimization-Based methods architecture

The learning scheme of analog circuit design is now composed by the interaction of two computational machines, the GA search optimizer and the SVM learning engine. Fig. 4.11 illustrates the block diagram for the optimization kernel with learning algorithm. The SVM models can influence the overall evolutionary process efficiency in two ways.

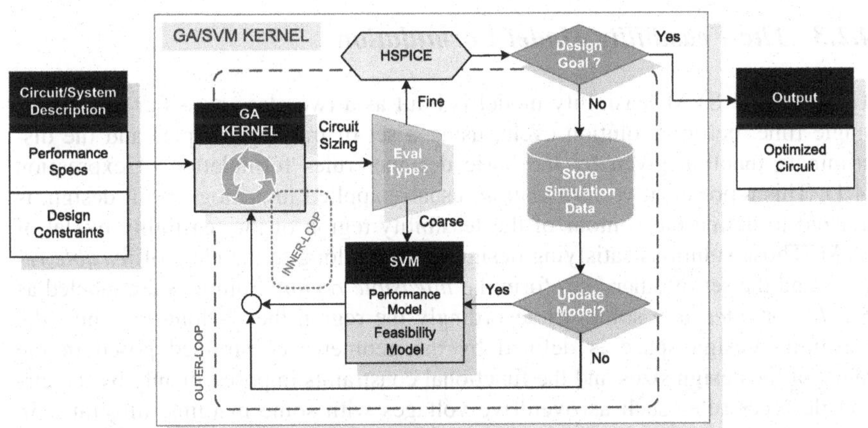

Fig. 4.11 Block diagram of the GA-SVM algorithm

When used as the performance model, the regression model establishes the mapping between the design variables and the performance parameters. This allows their combination to produce an approximation of the fitness function [20], as illustrated in Fig. 4.12, which is used to replace the expensive SPICE-like evaluations in the GA cycle. Potentially, this approach decreases the number of expensive true fitness evaluations and allows a better convergence rate.

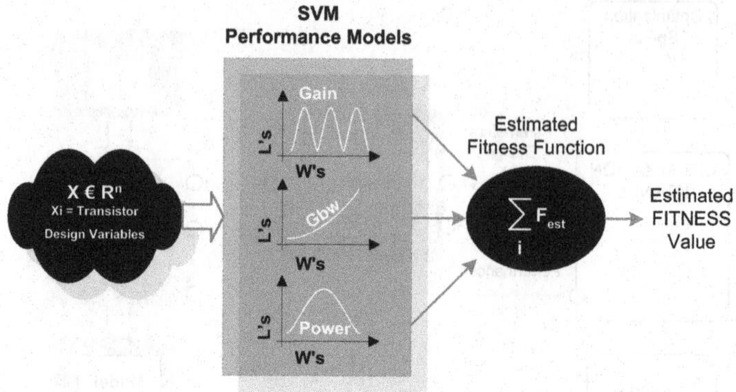

Fig. 4.12 Estimated fitness function with SVM performance model

However, the SVM model presented throughout this work is defined as a two class classifier model. The objective is to estimate the most promising regions, from the design space, to be explored. With this knowledge, the selection method will decide those solutions that will be accepted to proceed on the evaluation process, and those that will be rejected, because they are out or far from one of the most promising regions. The gain in this case is the number of avoided fine evaluations (normally heavy time-consuming electrical simulations) in each generation.

4.2.3 The Feasibility Model Formulation

The GENOM SVM feasibility model is built as a two class classifier model one single time, before evolution cycle, using a set of training samples and the discriminate function given by the basic designer rules formulation of expression (4.1). This representative formulation, usually applied in analog circuit design, is utilized to define the contour of the feasibility region of the feasibility model of SVM. Those solutions satisfying designer rules belong to the class of *feasible region*, and the set of other ones form the *infeasible region*. Solutions are labeled as *feasible* or *infeasible* solutions accordingly the region they belong to. Thus, the feasibility design space is defined by the geometry constrained posed in the range of the design sizes and the functional constraints imposed mainly by the circuit designer rules such as overdrive voltages with some margins, illustrated in Fig. 4.13.

$$VGS > VT + 50mV \quad and \quad VDS > VDSAT + 50mV \qquad (4.1)$$

To illustrate this concept in R^2, let us consider a simple Active RC low pass filter with gain *A0*, and frequency *f0* illustrated in Fig. 4.14. The feasibility contour is drawn with respect to capacitor *C1* an *R2*.

The feasibility model $M_f(x)$ defines a function that estimates the front-end between the feasible and infeasible space delimited by the geometry domain. The feasible sub-space, normally a small percentage of the total search space, is built

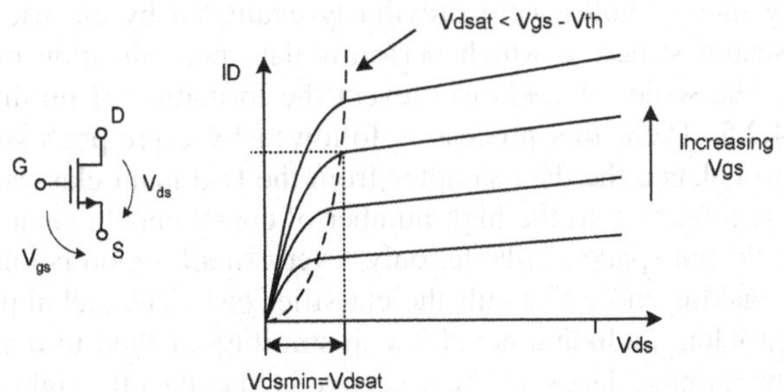

Fig. 4.13 Ids-Vds characteristic of short channel NMOS transistor

by the next sequence of actions. Taking the geometry constraints of the problem, each variable range is divided in equidistant points and is then evaluated by the circuit simulator.

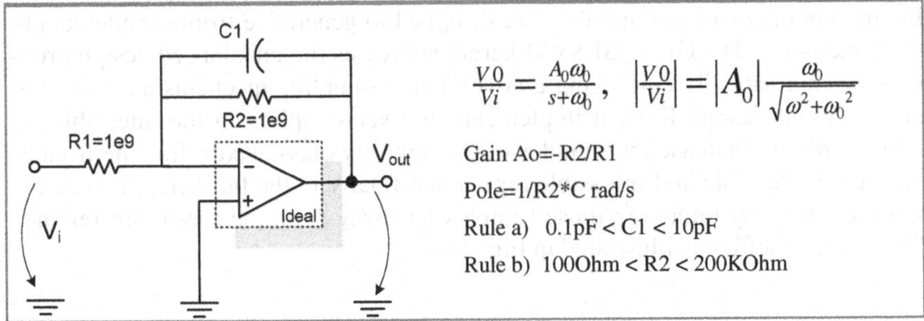

Fig. 4.14 Active RC filter

Another alternative is to sample a number of points proportional to the size of the design space. Next, they are classified in two data sets, the feasible data that satisfies all the designer rules and infeasible data, data which does not satisfy the design rules or was derived from convergence problems. Then, the samples are used as the train sequence to obtain the SVM classification model. The same HSPICE simulations used to build the feasibility model were reused to get the performance measures to train and build the SVM performance model for each performance parameter.

4.2.4 SVM Model Generation and Improvement

In order to improve the success and performance of the SVM feasibility model two enhancements were included in the model, a data sampling with parameter normalization preceded by an unbalanced data management mechanism.

The training data samples were previously evaluated by electric simulations, using a grid search structure which performs data normalization on the design variables using the scale [-1…+1] to prevent the formation of biasing models as explained in 4.2.5. Then, this process is followed by a pre-processing handling phase aiming to balance the data samples from the two main classes, the feasible and infeasible region. Due to the high number of constraints in analog design circuits and large design space available, only a very small region belongs to the interesting class making more difficult the classifier task. The techniques proposed to handle the problem include a novel 3-step stratified method to oversample and undersample the training data set. The objective is to collect the right subset of data samples from the pool of evaluated grid samples in order to build an efficient and accurate feasibility model. The implementation details are given next.

4.2.5 Handling Unbalanced Data in Circuit Designs

Unbalanced data problems impose some difficulties to the classifier task [12],[21-22]. The main pointed reasons are that most current classifier systems like SVM tend to optimize the overall accuracy without considering the weight of relative distribution of each class and they are designed to generalize from sample data to avoid the noise. The GENOM SVM kernel addresses the unbalanced design problems by automatically employing a novel 3-step sampling mechanism adjusted to analog circuit design. First, it implements an over-sampling in the infeasible region in order to increase the samples of the minority class, next refine the frontier between the feasible and infeasible region and finally in the third step, reduce the majority class, by removing those samples far away from the feasibility regions. The estimated effect is illustrated in Fig. 4.15.

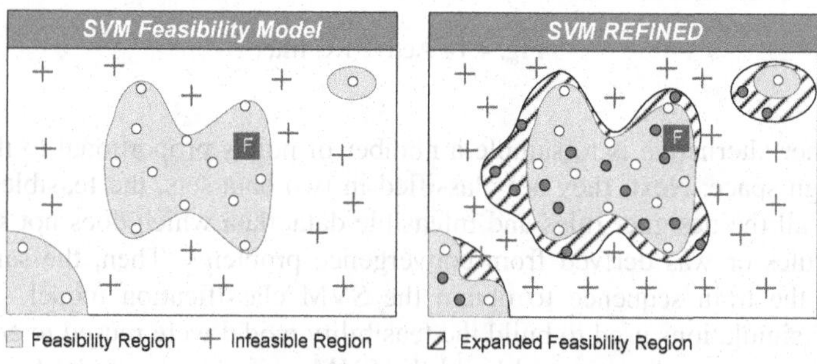

Fig. 4.15 Expected balance effect in design space

To accomplish these tasks, a first sampling strategy based on the classical grid search method as described in 3.3.2 is applied in first place. By default, during the evaluation phase a lot of statistics information is collected for each sample data among which the number of positive and negative samples for each class, the number of constraints satisfied given by the designer rules satisfied for each sample, the measure of constraints violation and so on. In addition, three new data sets were created: *Fs* embraces the set of evaluated samples that satisfies all constraints (feasible region), *Bs* appends the subset of sample data in boarder region (satisfies all constraints except one, two or three) and finally, *Is* attaches the remaining sample data in the infeasible region and sorts in ascending order of constraint violation value. Fig. 4.16, illustrates the idea of the search space subdivision into feasible and infeasibility regions.

The design experience acquired during this research, in several case studies, has shown that the ratio between positives and negatives samples is in order 0.04 to 0.07 for a total of 2000 uniform random points. An attempt to build a SVM feasibility model under such unbalanced data should result in low and biased performance models. To improve the estimation rate of the feasibility regions in new data, and to increase the efficiency of the model for more complex problems, two new sampling strategies were applied. First, oversampling the data of set *Fs* and Bs by random mutation in vicinity of the original data ("ball" mutation) and second, undersampling the elements of set *Is* by a factor equal to the unbalanced ratio, discarding always the last samples of the set.

Then, in the third step, a balanced SVM two class classifier model is finally built with the train data set, *Ts* being the union of the three final sets *Fs*, *Bs* and *Is* ($Ts = Fs \cup Bs \cup Is$) where the set of positive samples is given by *Fs*+*Bs* and *Is* is associated with the negative set. After that, the model is used to further generate new interior points of the feasibility region and neighborhood. Only the samples classified as positives will be evaluated by the true fitness function. In the end, the model is updated for the last time and the job is returned to the main process where it will be used together with an evolutionary algorithm to find the solution to the analog circuit design problem.

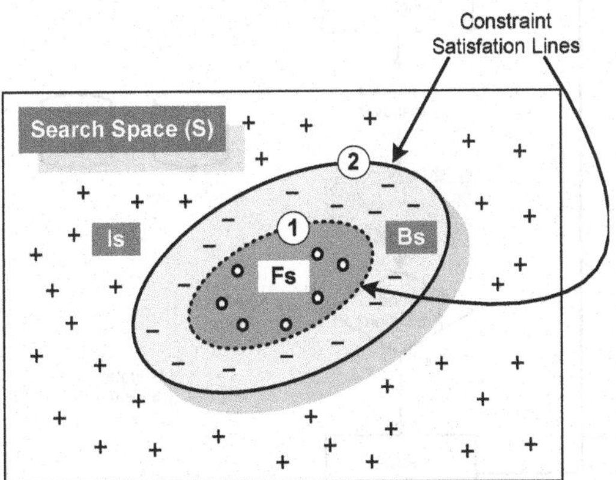

Fig. 4.16 Stratified vision of the search space by feasibility regions

4.2.6 GA-SVM Optimization Overview

The new GA-SVM approach uses all information acquired from the evolutionary process in order to make predictions about new data and improve the efficiency of the search algorithm.

The initialization phase of the GA algorithm is replaced by the sampling mechanism and model generation described in the two earlier sub-sections. Then, the evolutionary algorithm follows the sequential GA optimization algorithm with the exception of the evaluation phase. Here, the evaluation phase is preceded by an *active learning phase,* which uses the feasibility information from the model to decide which of the new offspring will be accepted, to proceed on the evaluation process and those that will be rejected from evaluation because they are out or far away from the most promising regions. The present approach uses a heavy time-consuming electrical simulator to evaluate the true fitness function for each submitted chromosome. Thus, the number of avoided fine evaluations identified by the active learning module in each generation represents a gain of efficiency of this approach and justifies one of the requirements of this implementation. The active evaluation process also implements an aggressive local search around the best individuals in the population, when the number of individuals selected by the active learning module is low. The block diagram of GA-SVM algorithm is illustrated in Fig. 4.17.

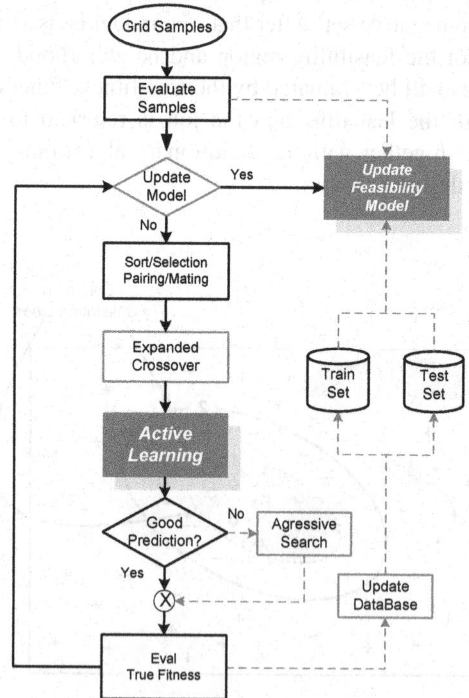

Fig. 4.17 Data flow of GA-SVM algorithm

4.2.7 Comments on the Methodology

This section clarifies the options taken by the presented methodology. To begin with, the constraint stratified vision (Fig. 4.16) used to deal with the unbalanced data problem of analog circuits design applications was implemented to pursue one the SVM fundamental principles, which says that only the support vectors contribute to the decision rule. To build a SVM model efficiently, using the constraint satisfied approach, it is only needed to manage the minority class set and the "best" infeasible samples because that is where the support vectors are present. This way, the management of the huge set of infeasible samples (majority class set) became simplified.

The choice of the training samples and the necessary initial grid resolution used to generate the SVM model, has a great impact on the quality of the model, and will affect the final model performance for unknown data. In the lack of a universal answer to this question, the approach taken in this research follows a simple rule, based on the percentage of the total search space and on the following belief: the feasibility model embodies the circuit's operational zone not in a single but in several points, satisfying or not the problem's specifications. The specs do not affect circuit's feasibility. In the first global sampling, the grid resolution should be chosen in such a way that, at least, one or several feasible points for each operational region should be detected or at least a reasonable number of positive sample points should be collected. If this condition is not met, the following measures can be taken: (a) increase the number of samples, by default it has the same effect of increasing the sampling resolution; (b) relax the contour of the feasible region, that is, accept in the feasible region those infeasible samples close to the feasible region; and/or (c) relax the constraints of the problem, this case needs user involvement.

4.3 Conclusions

The requirements of modern analog design automation tools are placing an increasing emphasis on analytic capabilities. Data mining technology has become an essential instrument in the analysis of large volumes of data in several activity domains. This chapter reviewed a SVM learning machine implementation applied to analog circuit optimization. SVM is considered one of the most efficient techniques belonging to the class of machine learning algorithm able to infer knowledge from data samples. This knowledge is useful to make predictions about new data or to get a better understanding of the system that generated the data. However, to manipulate an SVM tool with an acceptable level of usability and performance, four main tasks should be addressed: data normalization, data balancing, optimal parameters selection and data validation. The influence of these design decisions were illustrated by well known examples.

References

[1] Mitchell, T.: Machine learning. McGraw-Hill, New York (1997)
[2] Edelstein, H.A.: Introduction to data mining and knowledge discovery. In: Two Crows Corporation, 3rd edn. (2003), http://www.twocrows.com/intro-dm.pdf (Accessed March 2009)
[3] Anderson, D., Neil, G.M.: Artificial neural networks technology - A DACS state-of-the-art report. Tech Rep. Data and Analysis Centre for Software (1992)
[4] Jain, A.K., Mao, J., Mohiuddin, K.M.: Artificial neural networks: A tutorial. IEEE Computer, 31–44 (1996)
[5] Kroose, B.J., Smagt, P.V.: An introduction to neural networks. University of Amsterdam (1993), http://citeseer.ist.psu.edu/kroose93introduction.html (Accessed March 2009)
[6] Moore, A.W.: Regression and classification with neural networks. School of Computer Science Carnegie Mellon University (2006), http://www.cs.cmu.edu/~awm (Accessed March 2009)
[7] Farina, M.: A neural network based generalized response surface multiobjective evolutionary algorithm. In: Proc. Congress on Evolutionary Computation, vol. 1, pp. 956–961 (2002)
[8] Gaspar-Cunha, A., Vieira, A.: A Multi-Objective Evolutionary Algorithm Using Neural Networks To Approximate Fitness Evaluations. International Journal of Computers, Systems and Signals 6(1), 18–36 (2005)
[9] Burges, C.J.: A tutorial on support vector machines for pattern recognition. Data Mining and Knowledge Discovery 2(2), 121–167 (1998) http://www.umiacs.umd.edu/~joseph/ (Accessed March 2009)
[10] Moore, A.: K-means and hierarchical clustering - tutorial slides. School of Computer Science Carnegie Mellon University (2006), http://www.autonlab.org/tutorials/kmeans.html (Accessed March 2009)
[11] Nilsson, N.J.: Introduction to machine learning - an early draft of a proposed textbook. Department of Computer Science. Stanford University, Stanford (1996) http://robotics.stanford.edu/~nilsson/ (Accessed March 2009)
[12] Milenova, B.L., Yarmus, J.S., Campos, M.M.: SVM in oracle database 10g: Removing the barriers to widespread adoption of support vector machines. In: Proc. 31st International Conference on Very Large Data Bases, pp. 1152–1163 (2005)
[13] Smedt, B., Gielen, G.: WATSON: Design space boundary exploration and model generation for analog and RFIC design. IEEE Trans. Computer-Aided Design of Integrated Circuits and Systems 22(2), 213–224 (2003)
[14] Zizala, S., Eckmuller, J., Graeb, H.: Fast calculation of analog circuits feasibility regions by low level functional measures. In: Proc. Int. Conf. on Electronics, Circuits and Systems, pp. 85–88 (1998)
[15] Gräb, H., Zizala, S., Eckmüller, J., Antreich, K.: The sizing rules method for analog inte-grated circuit design. In: IEEE/ACM International Conference on Computer-Aided Design, pp. 343–349 (2001) doi:10.1109/ICCAD.2001.968645
[16] Wolfe, G.A.: Performance macro-modeling techniques for fast analog circuit synthesis. Ph.D. dissertation, Dept. of Electrical and Computer Engineering and Computer Science, College of Engineering, University of Cincinnati, USA (1999)

[17] Schwencker, R., Eckmueller, J., Graeb, H., Antreich, K.: Automating the sizing of analog CMOS circuits by consideration of structural constraints. In: Proc. Design, Automation and Test in Europe Conference and Exhibition, pp. 323–327 (1999)
[18] Nye, W., Riley, D.C., Sangiovanni-Vincentelli, A., Tits, A.L.: DELIGHT.SPICE: An optimization-based system for the design of integrated circuits. IEEE Trans. Computer-Aided Design 7(4), 501–519 (1998)
[19] Medeiro, F., et al.: A Statistical optimization-based approach for automated sizing of analog cells. In: Proc. ACM/IEEE Int. Conf. Computer-Aided Design, pp. 594–597 (1994)
[20] Kiely, T., Gielen, G.: Performance modeling of analog integrated circuits using least-squares support vector machines. In: Proc. Design, Automation and Test in Europe Conference and Exhibition, vol. 1, pp. 448–453 (2004)
[21] Liu, A.: The effect of oversampling and undersampling on classifying imbalanced text datasets, Master thesis, University of Texas, USA (2004)
[22] Romano, R.A., Aragon, C.R., Ding, C.: Supernova recognition using support vector machines. In: Proc. 5th International Conference on Machine Learning and Applications, pp. 77–82 (2006)

5 Analog IC Design Environment Architecture

Abstract. This chapter describes the implementation of an innovative design automation tool, GENOM which explores the potentials of evolutionary computation techniques and state-of-the-art modeling techniques presented in the previous chapters. The main design options of the proposed approach will be here described and justified. First, an overview of the design architecture main building blocks will be provided. Then, the optimization algorithm kernel, as well as, the implemented functionalities are described. Finally, the design options are described in detail using experimental results on a few test cases.

5.1 AIDA Architecture

The GENOM optimization tool can be used as a standalone application, although it holds some functionality which can only be fully accomplished when it is part of the in-house design automation environment called AIDA [1]. AIDA, Analog Integrated Design Automation, is an ongoing project for analog IC design automation at ICSG group IT/IST. A summary of this application architecture will be described next.

5.1.1 AIDA In-House Design Environment Overview

The AIDA platform, which includes a design flow core engine responsible for the design automation is illustrated in Fig. 5.1. The platform is structured in three layers: interface, application and data layer and implemented in several technologies, such as JAVA® for the design core, MySQL® for the databases and Swing® for the graphical user interface (GUI). The AIDA project implements a fully configurable design flow which introduces an increased level of flexibility and reusability when compared to traditional design approaches. The flexibility is achieved by both allowing the designer to define his own hierarchical design organization and, simultaneously, the design flow for each design. The reusability is achieved by introducing a highly organized data structure to store the entire design data allowing an easy reuse and retargeting of pre-design systems and predefined design flows. In addition, AIDA allows the interaction with other CAD tools such as circuit and system level optimizers like GENOM and layout generators [2-3].

M.F.M. Barros et al.: Analog Circuits and Systems Optimization, SCI 294, pp. 109–137.
springerlink.com

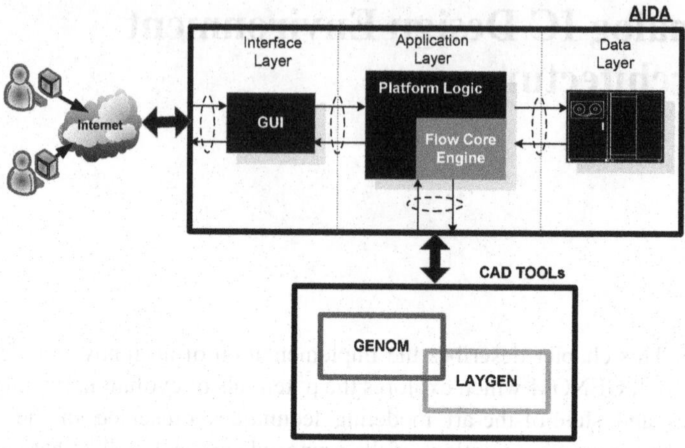

Fig. 5.1 Conceptual view of AIDA environment architecture

The AIDA platform implements a hierarchical methodology matching designers' approach by allowing the complete definition of the design flow tasks at each hierarchical level, as presented in Fig. 5.2 for a filter design case. The design flow definition is based on basic units of work: project specifications, topology selection, several units for device sizing and optimization and a last unit for characterization. In this project, GENOM acts as an external circuit and system level optimizer tool with well defined interface protocol.

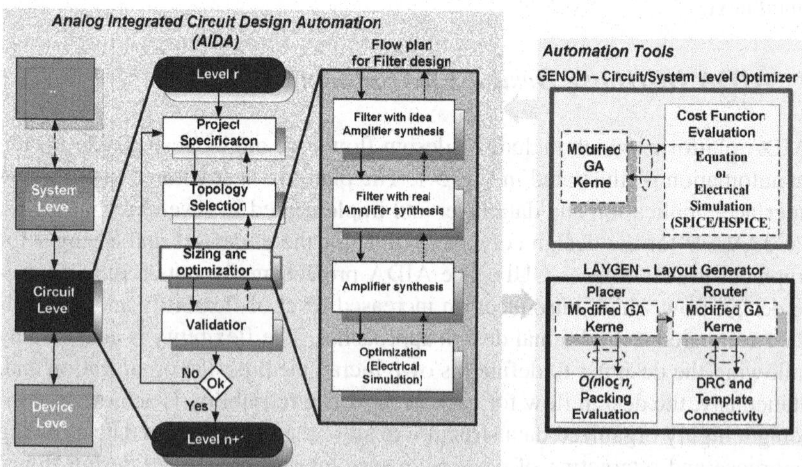

Fig. 5.2 AIDA design flow

The GUI facility of the AIDA platform, illustrated in Fig. 5.3, plays a key role in the definition of project specifications and topology selection required by GENOM.

Through an intuitive user-friendly interface, the user specifies the design specs e.g., circuit class, performance specs, design constraints and technology. These specs, which may be introduced by the user or result from the synthesis in a higher hierarchical level, automatically restrict the set of available topologies. Then, the topology selection may be performed manually by the designer or automatically by an engine (if available) that evaluates the candidate topologies according to design specs. Next, the design flow, organized in several design stages, controlling the optimization process, as exemplified in Fig. 5.2, is defined and executed. Each design stage has the goal of setting a subset of the design parameters (*W*, *L*, *C*, *R*, etc). Therefore, each design stage corresponds to an optimization task submitted to the selected optimization engine, in our case the GENOM optimization engine, using HSPICE, to compute the design objective function. Moreover, the use of other design and simulation tools, if available, is also possible and only depends on user's selection. Although a design stage is considered an atomic operation for the user, during the design flow and at each control point between design stages, he may evaluate the design and redefine parameters, constraints or even change the predefined design flow.

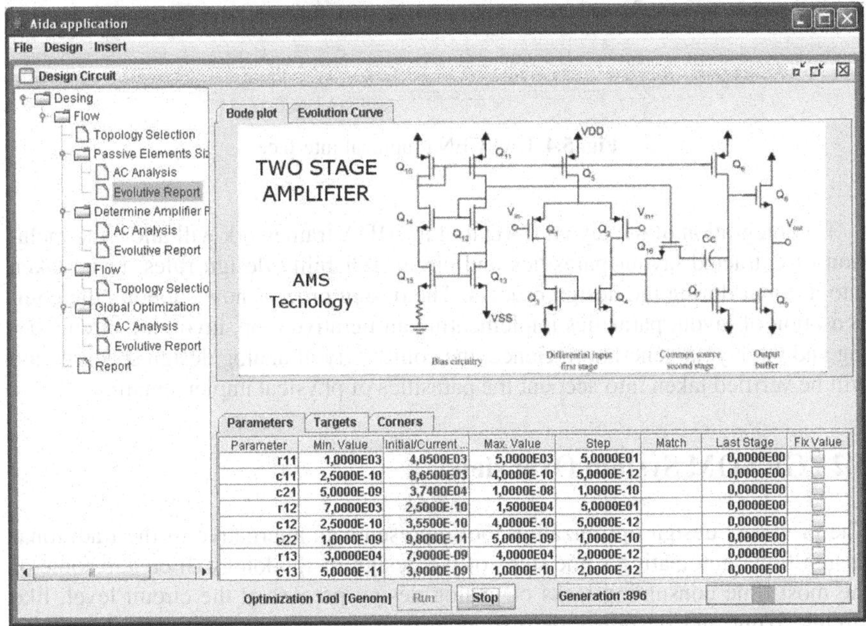

Fig. 5.3 GUI facility implemented in AIDA

5.1.2 Layout Level Tools

The AIDA framework was designed to interact with CAD tools of different hierarchical levels as described in the preceding section for the case of the analog circuit optimizer. In the future, this interaction will be expanded to the layout level for the layout verification and generation. In particular, the objective is to integrate the LAYGEN [2-3] tool illustrated in Fig. 5.4.

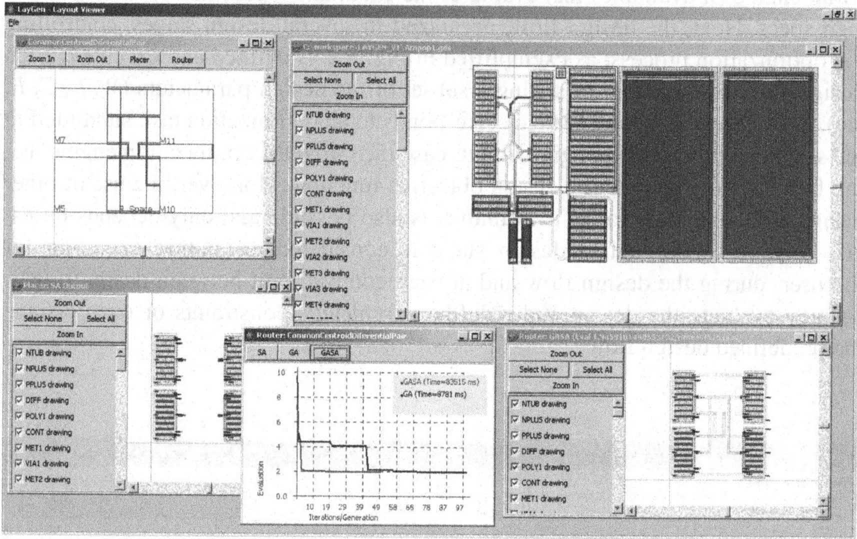

Fig. 5.4. LAYGEN graphical interface

The integration of the layout CAD tool in AIDA framework will allow the inclusion of extracted layout parasitics and circuit reliability design rules, to be taken into account during the design process. The design process now supports the compensation of layout parasitics implementing an iterative loop, involving circuit sizing and the layout generation. Hence, the conformity of analog design specification will be verified taken into account the parasitics of physical implementation.

5.2 GENOM System Overview

The proposed design optimization tool represents an alternative to the traditional design flow, automating some steps of the design methodology. It covers some of the most time consuming tasks of analog design process at the circuit level, like circuit sizing and design trade-offs identification. The main building blocks of GENOM architecture depicted in Fig. 5.5 are decomposed into three units, the optimization kernel, the evaluation module and the application interface.

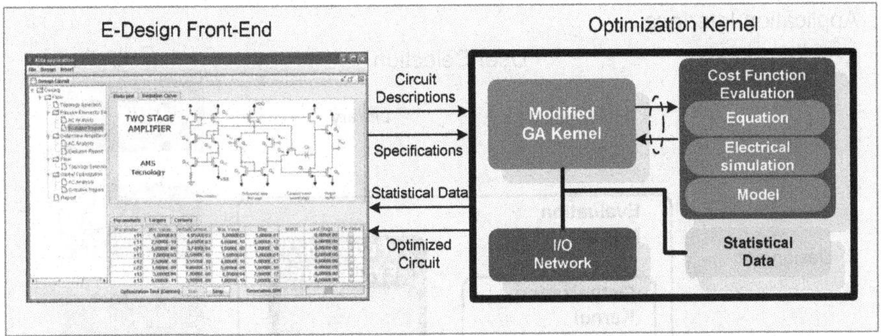

Fig. 5.5 E-Design environment architecture

The GENOM optimizer kernel is based on an evolutionary algorithm (EA) kernel with modified operators and an automatic control mechanism which supports the interaction with equation and simulation evaluation engines, so that the cost function evaluation is made either by behavioral models based on SVM or by electrical simulation, in this case, using Spice-like simulators. Additionally, GENOM includes a distributed processing facility with a high degree of portability across a variety of machines, allowing the increase in computation efficiency when using cost expensive evaluations.

The GENOM core is written in C, programming language, and implemented in a Linux environment, taking advantage of the efficiency and flexibility of C code, free development tools and platform. Although it is commonly used for algorithm development, C language has not traditionally been used to generate a graphical user interface (GUI) for applications. Hence, the front-end was implemented in an independent language platform, the Java™ using the Swing components.

The tool functionality, extended by the addition of an E-Design front-end allowing an incremental growth of the IC design database and an individual management of each project, will be described in the next sub-sections.

5.2.1 Design Flow

In order to support the analog IC design flow methodology and to provide an efficient data management of the inputs and outputs from GENOM, a new design automation environment was developed as illustrated in Fig. 5.5. Like in many analog design automation environments, before the synthesis there is a preparatory stage where the production of user-defined equations (equation-based), training the learning machine for performance models, or incorporating design constraints take place. The design facilities also include the backtracking of the design process, allowing the user to follow the evolution of the design process dynamically or just reporting the final solutions at the end of the optimization process. This feedback is extremely relevant once it provides the information that the designer needs to detect, identify and understand which are the performance bottlenecks for the circuit that is being designed.

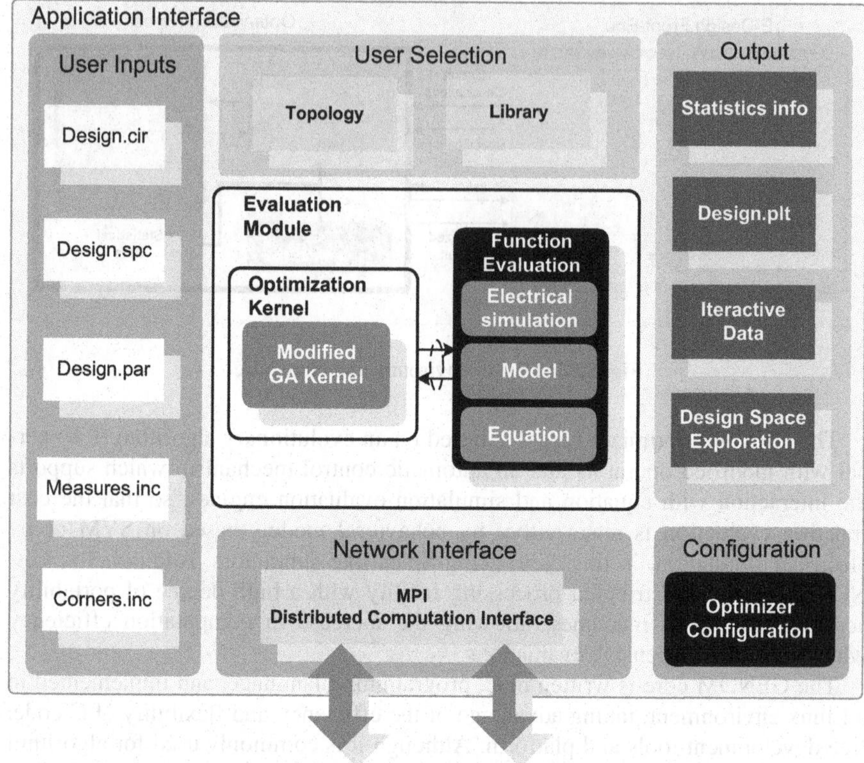

Fig. 5.6 Conceptual view of the Input/Output from optimizer tool

However, when not integrated in the AIDA environment, i.e., in the standalone operation, the user needs to provide and configure manually the necessary input files, depicted in Fig. 5.6, in a suitable form for the optimization process.

5.2.2 Input Data

The aim of this phase is to provide and configure the necessary input files in a suitable form for the optimization process. In order to manage the complex structure of data involved in this project, a graphical interface seems a fairly option to guide all the input data process. The GUI interface, using spreadsheet-like data input forms, aid the designer to input data more easily, minimizing input errors and the setup time to define or redesign a new simulation strategy. In addition, it guides the user through a sequence of logic events and avoids the occurrence of compatibility errors. Through the graphical input interface the user defines the circuit class (amplifier, filter, A/D, D/A, etc), the performance specs (dc gain, gain bandwidth product, phase margin, slew rate, power dissipation, offset voltages, etc) of the analog cell which the designer wants to optimize, as well as the design constraints (corners, matching parameters, overdrive voltages and currents, etc) and the technology process.

Fig. 5.7 illustrates one stage of design specs introduced by the user; in this case it shows the definition of the performance measures required for this project. According to the introduced design specs, a candidate topology is manually selected from the circuit database as depicted in Fig. 5.8. If the design specs do not match any of the existent topologies, a new one have to be created and introduced into the system.

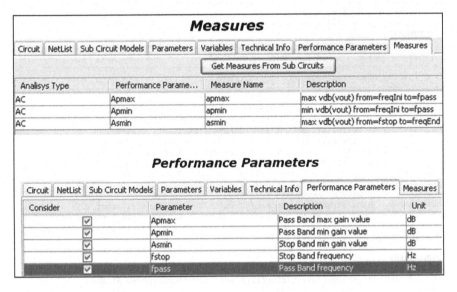

Measures

Circuit | NetList | Sub Circuit Models | Parameters | Variables | Technical Info | Performance Parameters | Measures

Get Measures From Sub Circuits

Analisys Type	Performance Parame...	Measure Name	Description
AC	Apmax	apmax	max vdb(vout) from=freqIni to=fpass
AC	Apmin	apmin	min vdb(vout) from=freqIni to=fpass
AC	Asmin	asmin	max vdb(vout) from=fstop to=freqEnd

Performance Parameters

Circuit | NetList | Sub Circuit Models | Parameters | Variables | Technical Info | Performance Parameters | Measures

Consider	Parameter	Description	Unit
☑	Apmax	Pass Band max gain value	dB
☑	Apmin	Pass Band min gain value	dB
☑	Asmin	Stop Band min gain value	dB
☑	fstop	Stop Band frequency	Hz
☑	fpass	Pass Band frequency	Hz

Fig. 5.7 Performance parameters and measures facilities

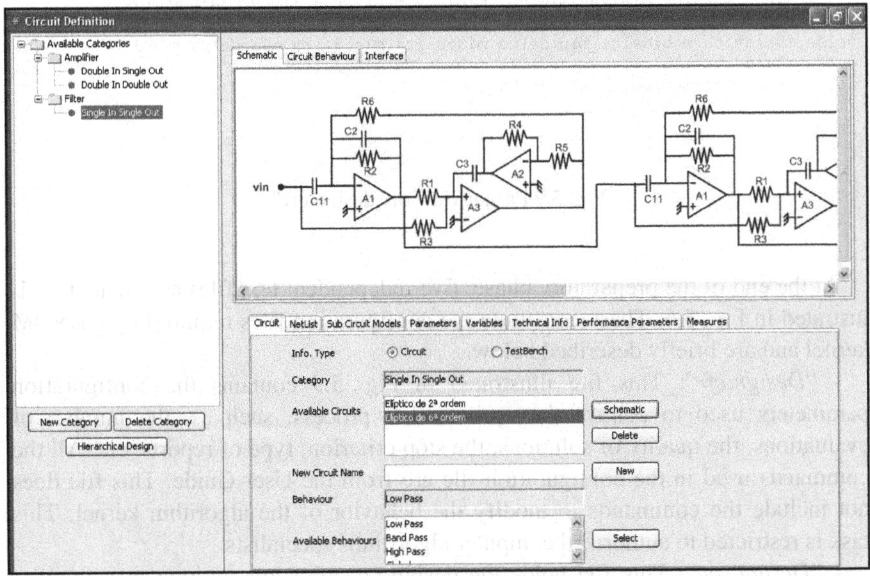

Fig. 5.8 Topology Selection

<design_file>**.cfr** - Configuration file

```
# A line started by a charater "#" is a comment.
<TITLE>
Differential AmPop
Version: November 16, 2007 - Author: F.M. Barros
# ---------------------------------------------------------------------------
# 1. Control Parameter Section
# ---------------------------------------------------------------------------
<CONTROL>
ProblemType      0     # 0 - Circuit simulation      1- Numerical optimization
OptimizationType 0     # 0 - Genetic algorithms      1- SVM (SA, ...)
# ---------------------------------------------------------------------------
# 2-Passing Parameter Section
# ---------------------------------------------------------------------------
<PASSING_PARAMETERS>
Seed     99  # SEED - Integer number representing the SEED value ={1-10000}
Timer     2  # TIMER- Simulation time TIMER={SHORT=0, MEDIUM=1, LONG=2}
Quality   2  # Optimization QUALITY={COARSE=0, MEDIUM=1, FINE=2}
Stop      2  # STOP Criterion. STOP={Time=0, Convergence=1, Max_Generations=2}
Debug     1  # DEBUG     - Output text debugging. DEBUG={none=0, YES=1 }
Cluster   0  # CLUSTERS - Parallel Processing ={SERIE=0, PARALLEL=1}
Reports   0  # REPORTS   - Formats {TEXT=0, GRAPHICS=1, Both=2}
Activity 10  # ACTIVITY - Statistics data sampling frequency (for graphics)
StepAC   10  # STEPAC    - Update frequency of bode plots
inDirectory /home/IT/GENOM/workspace/circuits/00_Differential_Ampop
outDirectory /home/IT/GENOM/workspace/circuits/00_Differential_Ampop/RESULTS
# ---------------------------------------------------------------------------
# 3-Dependent Parameters Section
# ---------------------------------------------------------------------------
<MEASURES>
9
gain_dc;gbw;phfp;phase_margin;ftcmfb;phfpcmfb;phasecmfb;power_a;iavdd
##
<CONSTRAINTS>
34
vov_m0a;vov_m0b;vov_m16;vov_m1a;vov_m1b;vov_m2a;vov_m2b;vov_m3a;vov_m3b
vov_m4a;vov_m4b;vov_m5a;vov_m5b;vov_m6a;vov_m6b;vov_m7a;vov_m7b
delta_m0a;delta_m0b;delta_m16;delta_m1a;delta_m1b;delta_m2a;delta_m2b
delta_m3a;delta_m3b;delta_m4a;delta_m4b;delta_m5a;delta_m5b
delta_m6a;delta_m6b;delta_m7a;delta_m7b
...
```

Fig. 5.9 Partial view of "design.cfr"

At the end of the preparatory phase, five independent text files are created as illustrated in Fig. 5.6. These constitute the configuration files required by GENOM kernel and are briefly described below.

- "*Design.cfr*": This file illustrated in Fig. 5.9 contains the configuration parameters used to control the optimization process, such as, the number of evaluations, the quality of solutions, the stop criterion, type of reports, etc. All the commands used in the configuration file are from the User Guide. This file does not include the commands to modify the behavior of the algorithm kernel. This task is restricted to authorized computer algorithms specialists.

- "*Design.spc*": This file holds the design specifications written in a familiar analog design syntax, using the traditional relational "*min*", "*max*", "less", "*great*", "*equal*" operators and additional ones for specific constraints expressions such as "*verify_bound(a,b,c)*" illustrated in Fig. 5.10.

- "*Design.par*": The design parameters file depicted in Fig. 5.11 encloses the problem dimension and device names, bounds and step size for each optimization variable.

- "*Design.cir*": This is the circuit netlist file that describes the circuit connectivity either in flattened or hierarchical mode. The optimization variables must be explicit marked with an underscore before the variable's name as depicted in Fig. 5.12. This name must agree with at least one parameter of the design parameters file. The format of this file should be compatible with the evaluation tool.

- "*Corners.inc*": This is an optional input file that specifies the corners conditions. This file showed in Fig. 5.13, will be included in the circuit netlist.

- "*Measures.inc*": This is a user-defined set of statements or commands that retrieve specific electrical measures from evaluation tool. It is a kind of interface between optimizer and the evaluation tool to acquire precise information data. This file, illustrated in Fig. 5.14, will be included in the circuit netlist.

- *Fabrication model*: A fabrication model consists of values for different transistor characteristics needed by the simulator to develop a small signal model for a transistor. In a regular basis, this file is complied with standards and is dependent on the fabrication technology. In GENOM, a library of models aggregates some of the public technological models available. The technological file must be referenced in <Design.cir> file, as illustrated in Fig. 5.15.

<design_file>**.spc** - Specs File

```
  (gain_dc > 55)
+ (gbw     > 100e6)
+ (verify_bound(phase_margin,60,90))
+ (ftcmfb  > 50e6)
+ (verify_bound(phasecmfb,60,90))
+ (min(power_a,0,10e-3))
+ (min(iavdd,0,10e-3))
+ (check_bound(vov_m0a,100e-3,300e-3))
+ (check_bound(vov_m0b,100e-3,300e-3))
+ (check_bound(vov_m1a,50e-3,300e-3))
+ (check_bound(vov_m1b,50e-3,300e-3))
+ (check_bound(vov_m2a,100e-3,300e-3))
+ (check_bound(vov_m2b,100e-3,300e-3))
...
+ (check_bound(delta_m0a,100e-3,1000))
+ (check_bound(delta_m0b,100e-3,1000))
+ (check_bound(delta_m1a,100e-3,1000))
+ (check_bound(delta_m1b,100e-3,1000))
+ (check_bound(delta_m2a,100e-3,1000))
+ (check_bound(delta_m2b,100e-3,1000))
...
```

Fig. 5.10 Partial view of <design.spc>

<design_file>**.par** - Optimization Parameters File

```
21                      # Number of optimization variables of the problem
_w00                    # 1st Device name
1.0e-6                  # Inferior bound and
300.0e-6                # Upper bound
1.0e-6                  # Step size

_L00                    # 2nd Device name
0.35e-6                 ...
10.0e-6                 ...
0.05e-6                 ...

_m01                    # 3rd Device name
1                       ...
100                     ...
1                       ...
...

_c10                    # 21st Device name
1.0e-12                 # Inferior bound and
100.0e-12               # Upper bound
1.0e-12                 # Step size
...
```

Fig. 5.11 Partial view of <design.par>

<design_file>**.cir** - Circuit Netlist

```
* Differential Ampop Revised: Monday, November 16, 2007
* D:\IT\GENOM\CIRCUITS\AMPOP\AMPOP.DSN       Revision: 1
...
M20 VB3 VB3 N07332 0 nmos   w=_w04 l=_104 m=1
M21 N07332 VB3 0 0 nmos     w=_w11 l=_111 m=1
M5A N02095 CMFB 0 0 nmos    w=_w04 l=_105 m=_m05
M5B N01845 CMFB 0 0 nmos    w=_w04 l=_105 m=_m05
M7A N11287 N11287 0 0 nmos  w=_w04 l=_105 m='_m05/2'
M7B CMFB N11287 0 0 nmos    w=_w04 l=_105 m='_m05/2'
*
M6A N11287 VCMI N10772 avdd pmos w=_w02 l=_106 m=_m06
M6B CMFB   VCM  N10772 avdd pmos w=_w02 l=_106 m=_m06
*
R10 VOUTP VCMO _R1
*
C10 VCMO VOUTP _c10
*
**********************
**** .DATA info  ******
**********************
.DATA PIPEdata LAM
FILE = 'PIPE.dat' _w00=1 _w01=2 _w02=3 _w04=4 _w10=5 _w11=6 _100=7 _101=8
_102=9 _103=10 _104=11 _105=12 _106=13 _110=14 _111=15 _m01=16 _m02=17 _m03=18
_m04=19 _m05=20 _m06=21 _R1=22 _c10=23
.ENDDATA

**************************
* setting for AC analysis
**************************
.ac dec 50 1 1e9 SWEEP DATA = PIPEdata
* plot data
.probe ac vdb(voutd) vdb(vcmo) vp(voutd) vp(vcmo)
```

Fig. 5.12 Partial view of <design.cir>

```
<Corners>.inc - Corners File (HSPICE style)

* ------------------------
* 1. Corners file
* ------------------------
.ALTER @1 -> lib=slow; temp=-40 +50 +105;
      .protect
      .lib 'cmos035.lib' slow
      .unprotect
      .temp -40 +50 +105

.ALTER @2 -> lib=typ; temp=-40 +50 +105;
      .protect
      .lib 'cmos035.lib' typ
      .unprotect
      .temp -40 +50 +105

.ALTER @ -> lib=typ; temp=-40 +50 +105;  ...
```

Fig. 5.13 Partial view of the Corners file

```
<Measures>.inc - Measures Files (HSPICE commands)

* The Measures Section
* ------------------------
* A. Measures for SPECs
* ------------------------
.MEASURE AC 'gain_dc' max vdb(voutd) from=1 to=1000
.MEASURE AC 'gbw' when vdb(voutd)=0 fall=1
.MEASURE AC 'phfp' find vp(voutd) at=gbw
.MEASURE AC 'phase_margin'     PARAM('phfp + 180')
.MEASURE AC 'ftcmfb' when vdb(vcmo)=0 fall=1
.MEASURE AC 'phfpcmfb' find vp(vcmo) at=ftcmfb
.MEASURE AC 'phasecmfb' PARAM('phfpcmfb+180')
.MEASURE AC 'power_a'   PARAM('-P(vdd)/2')
.MEASURE AC 'iavdd'     PARAM('-P(vdd)/avddpar/2')
* -----------------------------------------------
* B. Transistor Bias Measures - overdrive voltage
* -----------------------------------------------
.measure AC vov_m0a    = param('VGS(xampop.m0a)-VTH(xampop.m0a)')
.measure AC vov_m0b    = param('VGS(xampop.m0a)-VTH(xampop.m0b)')
.measure AC vov_m1a    = param('VGS(xampop.m1a)-VTH(xampop.m1a)')
.measure AC vov_m1b    = param('VGS(xampop.m1a)-VTH(xampop.m1b)')     ...

* -----------------------------------------------------
* C. Transistor Transistors Vds voltage margin to VDsat
* -----------------------------------------------------
.measure AC delta_m0a    = param('VDS(xampop.m0a)-VDSAT(xampop.m0a)')
.measure AC delta_m0b    = param('VDS(xampop.m0a)-VDSAT(xampop.m0b)')
.measure AC delta_m1a    = param('VDS(xampop.m1a)-VDSAT(xampop.m1a)')
.measure AC delta_m1b    = param('VDS(xampop.m1a)-VDSAT(xampop.m1b)') ...
...
```

Fig. 5.14 Partial view of the measures file

- Cost Function: This is a module that implements a parser in Lex and Yacc syntax [4] which automatically evaluate the performance of a set of candidate solutions. It is independent from the problem and will be the subject for further discussion in sub-section 5.3.3.1.

<Fabrication>**.inc** - Technology Process File (HSPICE style)

```
**************************
* Libs
**************************
.protect
    .lib '../../library/cmos035/cmos035.lib' typ/slow/fast
            or
    .lib '../../library/UMC/HSPICE/telescopic/l18u18v.122' L18U18V_TT
            or
    .lib '../../library/AMS/hspiceS/c35/wc49.lib' tm/wp/ws
.unprotect
```

Fig. 5.15 Technological model reference

5.2.3 *Output Data*

The output data provided by the GENOM tool includes the post-processing reports and evolutionary real time reports. The activation of each type of outputs is left to the designer choice. The post-processing reports include the evaluation of performance parameters coupled with statistical information presented at the end of the optimization, using the data in the data structures generated during the optimization phase. Fig. 5.16 and Fig. 5.17 illustrate the type of documentation provided by the design automation environment. The GENOM outputs are divided in two great groups related with design data and process info.

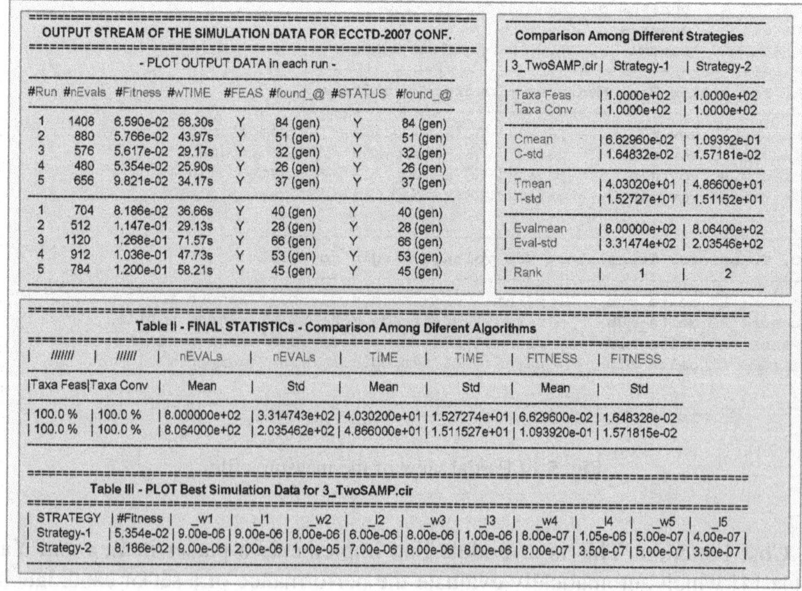

OUTPUT STREAM OF THE SIMULATION DATA FOR ECCTD-2007 CONF.

- PLOT OUTPUT DATA in each run -

#Run	#nEvals	#Fitness	#wTIME	#FEAS	#found_@	#STATUS	#found_@
1	1408	6.590e-02	68.30s	Y	84 (gen)	Y	84 (gen)
2	880	5.766e-02	43.97s	Y	51 (gen)	Y	51 (gen)
3	576	5.617e-02	29.17s	Y	32 (gen)	Y	32 (gen)
4	480	5.354e-02	25.90s	Y	26 (gen)	Y	26 (gen)
5	656	9.821e-02	34.17s	Y	37 (gen)	Y	37 (gen)
1	704	8.186e-02	36.66s	Y	40 (gen)	Y	40 (gen)
2	512	1.147e-01	29.13s	Y	28 (gen)	Y	28 (gen)
3	1120	1.268e-01	71.57s	Y	66 (gen)	Y	66 (gen)
4	912	1.036e-01	47.73s	Y	53 (gen)	Y	53 (gen)
5	784	1.200e-01	58.21s	Y	45 (gen)	Y	45 (gen)

Comparison Among Different Strategies

3_TwoSAMP.cir	Strategy-1	Strategy-2
Taxa Feas	1.0000e+02	1.0000e+02
Taxa Conv	1.0000e+02	1.0000e+02
Cmean	6.62960e-02	1.09392e-01
C-std	1.64832e-02	1.57181e-02
Tmean	4.03020e+01	4.86600e+01
T-std	1.52727e+01	1.51152e+01
Evalmean	8.00000e+02	8.06400e+02
Eval-std	3.31474e+02	2.03546e+02
Rank	1	2

Table II - FINAL STATISTICs - Comparison Among Diferent Algorithms

//////	/////	nEVALs	nEVALs	TIME	TIME	FITNESS	FITNESS
Taxa Feas	Taxa Conv	Mean	Std	Mean	Std	Mean	Std
100.0 %	100.0 %	8.000000e+02	3.314743e+02	4.030200e+01	1.527274e+01	6.629600e-02	1.648328e-02
100.0 %	100.0 %	8.064000e+02	2.035462e+02	4.866000e+01	1.511527e+01	1.093920e-01	1.571815e-02

Table III - PLOT Best Simulation Data for 3_TwoSAMP.cir

STRATEGY	#Fitness	_w1	_l1	_w2	_l2	_w3	_l3	_w4	_l4	_w5	_l5
Strategy-1	5.354e-02	9.00e-06	9.00e-06	8.00e-06	6.00e-06	8.00e-06	1.00e-06	8.00e-07	1.05e-06	5.00e-07	4.00e-07
Strategy-2	8.186e-02	9.00e-06	2.00e-06	1.00e-05	7.00e-06	8.00e-06	8.00e-06	8.00e-07	3.50e-07	5.00e-07	3.50e-07

Fig. 5.16 Progress reports

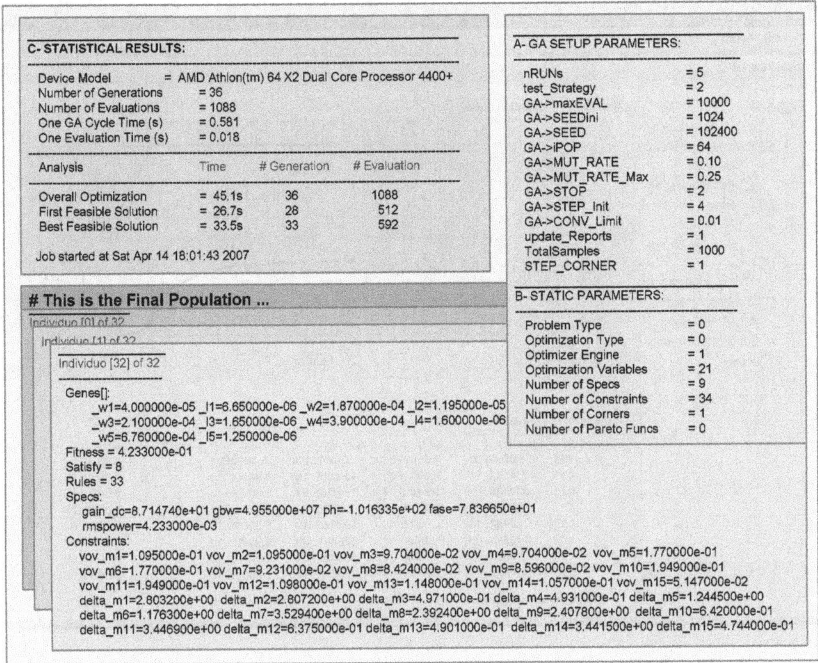

Fig. 5.17 Performance reports from optimization

Process info: This is the union of several statistical metrics gathered from optimization (Fig. 5.16). It includes a huge amount of statistics data about runs, generations, evaluations and time. This data is spread in several thematic files, including the evolution report file, corners file, bode plot file, etc. Optionally, the user can dispatch this info to screen reports for "online" validation purpose as it will be discussed in the next section.

Design Data: This corresponds to the final results from the estimation process (Fig. 5.17). This includes the optimum values of the optimization variables, the performance parameter values and the satisfaction of constraints parameters for the best 32 individuals of the population. In addition, it provides information about the optimization problem progress. These values are confronted with the initial ones to infer about the fulfillment of the synthesis flow objectives.

5.2.3.1 Progress Real-Time Reports

GENOM produces and supplies the required data which allows the visualization of real-time reports in AIDA framework. The progress real-time reports are a set of visual tools available optionally to the user, which indicate the progress status of evolutionary process in each generation. They consist of animated graphics of bode plot figures, the design space exploration figures and of the evolution curve of the cost function. The real-time environment is also represented by a

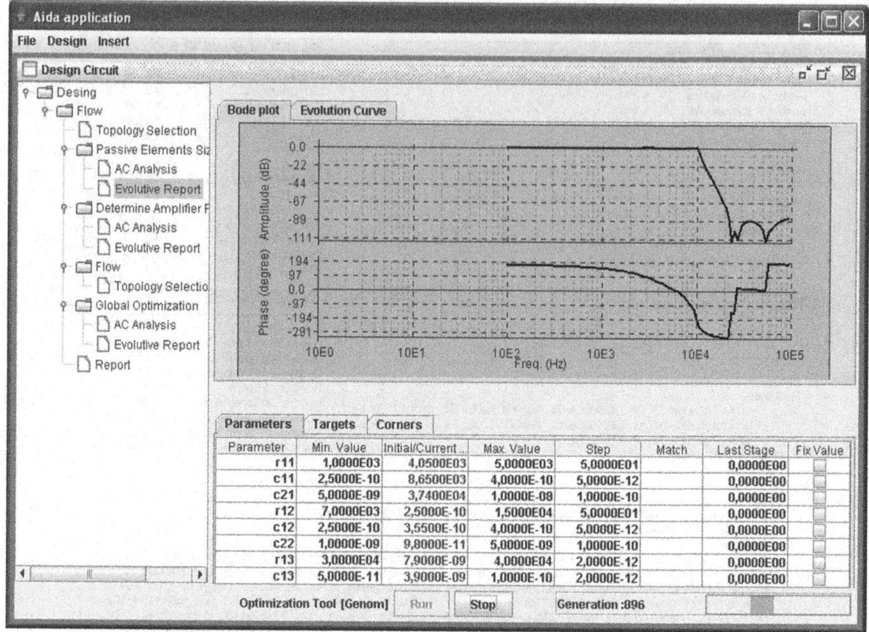

Fig. 5.18 Progress reports provided by the automation prototype

built-in spec sheet that can display a simple pass/fail status, symbolized by green/red colors, of the performance parameters, constraints violations and corners satisfaction as illustrated in Fig. 5.18.

5.2.3.2 Interactive Design

Interactive design is an extended capability introduced to GENOM framework that allows an experienced designer to incorporate some basic knowledge about a circuit during the search process. With the feedback acquired from real-time progress reports, for example, comparing the initial specs against current measured results and taking into account the present context status of the optimization process (state of design variables, evolution curve, constraints violation and corners satisfaction, etc.), the designer can use his knowledge or intuition to change the dynamic ranges of design parameters, set fixed values to genes of the current population (affix some genes of chromosome), etc, which shifts the course of optimization. Keeping constant values in some design variables has the effect of reducing the number of search variables. One equivalent variation of this approach is done by the matching of some strategic transistors such as, the differential pairs, current mirrors, etc., and in some non-sensitive transistors because they do not have much impact on the functionality of the circuit. Both measures result in the shrinkage of the design space and shortened run times. The advantage of this approach is that it is independent from the process, it captures the designer knowledge and since it adapts to each individual's knowledge, it is more flexible and can lead to efficient performances. Interactive design becomes a valuable optional tool in the presence of an experienced designer.

5.2.4 I/O Interfaces

The MPI interface block illustrated in Fig. 3.24 is composed by two independent types of communications. The hierarchical level interface is dedicated to future integration with LAYGEN tool. The network communication interface implements a local area multi-computer LAM-MPI interface (Fig. 5.19) used in the development of parallel applications over a network of heterogeneous computers as described in sub-Sect. 3.3.7.

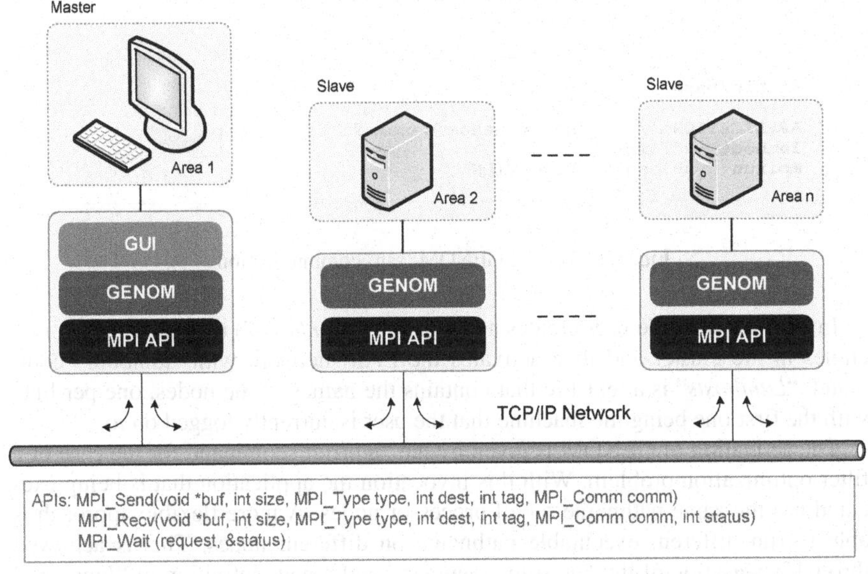

Fig. 5.19 Local area multi-computer system implemented with LAM-MPI

As discussed in 3.3.7, the communication between parallel processes is handled by the Message Passing Interface (MPI). Therefore, it is necessary to download, compile and install the MPI library in the current environment according to instructions in "GENOM Users guide". To make sure that distributed optimization environment is correctly configured and installed in a specific processing node, execute the "test-GENOM" script of Fig. 5.20:

```
#!/bin/bash
# test-GENOM - A script to test remote opt. environment
MACHINE_IP=$1
echo "GENOM: Testing remote machines"
echo "Testing remote GENOM is available …"
ssh  $MACHINE_IP genom
echo "Now, testing if remote simulator is available …"
ssh  $MACHINE_IP hspice
echo "Now, testing SSH to avoid passwords …"
ssh  -x -a $MACHINE -n echo $SHELL
```

Fig. 5.20 Testing GENOM distributed environment

The latter script verifies if the optimization tool, as well as, the evaluation engine are available in a specified processing node by trying to execute an application, e.g."*genom*" and "*hspice*", on all nodes. The last test verifies if the secure "SSH" communications is configured to avoid passwords. If the test is successful, proceed with next sequence of commands to initiate the execution of parallel application, the "*genom*" in the example illustrated in Fig. 5.21.

```
#!/bin/bash
# test-GENOM - A script to activate distributed processing
APPLICATION='/home/genom/Genom/genom filtro.cfr -p'
lamboot -v lamhosts
mpirun C sh.csh $APPLICATION
```

Fig. 5.21 Testing GENOM 'ssh' communications

In the first step, the user creates a file listing ("*lamhosts*") the participating machines in the cluster and then activates the LAM network with "*lamboot*" command. "*Lamhosts*" is a text file that contains the names of the nodes, one per line, with the first one being the machine that the user is currently logged on to.

The activation of GENOM is given by the "*mpirun*" command for the case of a filter optimization problem. With this invocation the application that is being executed has the same pathname on all processor nodes. A more flexible approach is able to run different executable pathname on different nodes. This is achieved through a variation of the "*mpirun*" command and a new definition of "*lamhosts*" as described in Fig. 5.22.

```
#!/bin/bash
# test-GENOM - A script to activate distributed processing
APPLICATION='/home/genom/Genom/genom filtro.cfr -p'
mpirun -p4pg <lamhosts> $APPLICATION
```

Fig. 5.22 Invocation of distributed GENOM application

For example, to run "*genom*" program on machine *baltar*, *malacata* and *everest* all Linux machines, and on *estrela*, a Solaris machine, the <lamhosts> file would contain now the following entries depicted in Fig. 5.23:

```
# a 4-node LAM running 5 processes
baltar    0 /home/PhD/Work/AIDA/GENOM/bin/genom filtro.cfr -p
malacata 2 /home/gneves/AIDA/GENOM/exe/genom filtro.cfr -p
everest  1 /usr/local/linux/GENOM/genom filtro.cfr -p
estrela  1 /home/ngonc/Solaris/GENOM/bin/genom filtro.cfr -p
```

Fig. 5.23 Lamhosts with the names of nodes and the pathname to the executable

The second entry per line, here 0, 2, 1 and 1, is the number of additional processes that can be launched per each machine. Since the MPI run is started from *baltar* the master process runs on it, so it is advisable not to allow the execution of another process on it. The other nodes have associated one or two processes per machine. This approach presents several advantages because it is possible to apply efficient load management of computer power in unbalanced network. An unbalanced network occurs when the computer power distribution is not equally distributed between machines, either due to different machines or to machines with different loads. Balancing the number of processes according to the available computational resources reduces the overall optimization time.

5.2.5 Evaluation Engine

GENOM extends the optimization capabilities to some of the SPICE-like circuit simulators including the standard HSPICE and SPICE which share common characteristics. These simulators are capable of reading their inputs and producing results in text file formats, as well as, being launched from the command line. Other simulators can also be supported as long as these characteristics occur. A detailed description of the entire mode of operation ranging from the moment a chromosome is ready for evaluation until it attains the cost function value is presented in section 5.3.3.

5.2.6 Expansion of GENOM Tool

The GENOM synthesis tool consists of a set of interconnected software modules which comprise the user interface, the evaluation engine, the distributing computing API and the learning machine beyond the optimization engine itself. These modules are called automatically when required by the synthesis flow. Preferentially, AIDA uses a XML text description files to pass information between internal modules taking advantage of the intrinsic XML properties. The XML file format provides the developer with a clean, robust and human readability documented target, allowing a much easier debugging as well as reading and exporting to other file formats. If the necessary software modules are developed, then the presented system can also be applied to different design environments or can even be integrated in wider industrial applications. Fig. 5.24 depicts an excerpt of the configuration interface file used by AIDA framework to setup some functionalities of the GENOM tool.

```
/*************************************************************************/
                    interface.c - configuration file
           Copyright (C) 2005 by Manuel Barros, fmbarros@ipt.pt
/*************************************************************************/
#   This file contains the INPUT parameters to GENOM Optimizer- V.2
#   Using the command line:
#                   Ex: ../genom RcIdeal.cfr -s -hspice
 *************************************************************************/

<?xml version="1.0"?>
<AIDA>
<GAPAR>           # Optimization GA Algorithms under test
<num of runs>      20          #number of runs
<evaluations_max> 10000       #number maximum of evaluations
<initial_seed>     1000        #initial seed
<population_size> 64          #population size
<mut_rate_max>     0.25        #maximum mutation rate
<stop_criteria>    1           #"1=Maximum num_generation 2=1st solution 3=25 STAGNATED generations"
<convergence_lim> 10e-3       #Cost standard deviation limit for the convergence test
<update_report>    1           #1 = each generation 2= logaritmic 3= best changed
<num_of_runs>      20          #number of runs
<update_report>    1           #update reports
<ntotalsamples>    3000        #number of total samples
</GAPAR>
#
<KERNEL> # Optimization Algorithms under test
 ...
 ...
</KERNEL>
</AIDA>
```

Fig. 5.24 Interface between GENOM and AIDA design automation environment.

The Fig. 5.25 demonstrates a communication interface example resultant from the <update_reports> parameter specification defined in Fig. 5.24. At specific time intervals pre-defined by the user, it is carried out an update of the reports and the refresh of screen information. In the example above, <update_reports> is set to '1' meaning an update in each generation (see Fig. 5.24 for other options). The information delivered from the optimization tool intended for visualization purposes is treated by a parser that identifies pairs of keywords or tags (*fSpecs.out*, *fEvolution_Curve.out*, *fCorners.out*, *fParameters.out*). The information between those keywords is sent to the interface defined by the client (the entity that initiated the optimization order). Fig. 5.25 exemplifies one line of results sent from GENOM. The word "*fSpec.out*" is reserved and identifies the performance parameters and the following values have a precise syntax. The first argument specifies the iteration of evolutionary algorithm and the next ones are the optimal values for the performance in the same order of appearance as in the specs file ("*design.spc*").

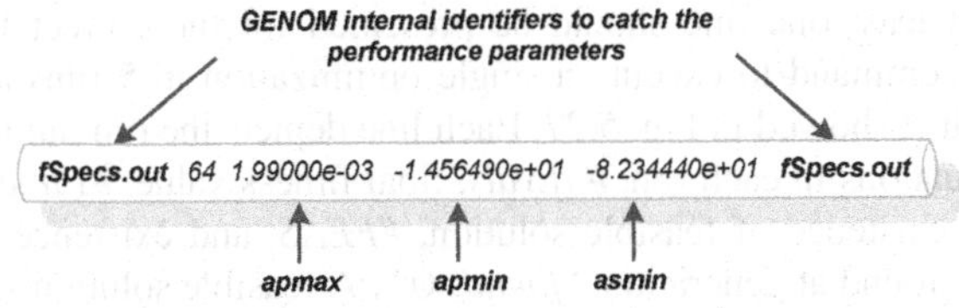

Fig. 5.25 Example of information delivered by GENOM

5.2.7 Optimization Kernel Configuration

This section presents the implemented approaches that support the optimization kernel. GENOM includes a kernel configuration file with commands to modify the behavior of the algorithm kernel. This task is limited to authorized computer algorithms specialists. Fig. 5.26 depicts a sample of the configuration interface file “AGPAR.h” used to setup some GENOM functionalities.

Each line between <KERNEL> tags is represented by a set of attributes that defines a particular characteristic of the kernel. The example, depicted in Fig. 5.26 defines the optimization of three different kernels, “GA-STD”, “GA-MOD” and

```
/************************************************************************/
                    AGPAR.h - configuration file
          Copyright (C) 2005 by Manuel Barros, fmbarros@ipt.pt
/************************************************************************/
#   This file contains the the commands used to modify the behavior of the
#   algorithm kernel. This file has restricted access.
 ************************************************************************/
<?xml version="1.0"?>
…
<COMMENTS>
1=Name             # "Name and specific GA PARAMETERS details"
2=strategy         # "1= Evolution Strategy  2=ES+SVM_REGRESSION 3=ES+SVM_CLASS 4= 1+2+3"
3=sampling         # "Initial Sampling [1= Random  2=DOE  3=LHS  4=outros]"
4=sort             # "Sort method 0= by cost function   1=by Feasibilidade"  2= by Constraints
5=selection_type  # "1= Random  2=Roulette wheel 3=Tournment Selection" 4= By Feasib. 5= By Masks"
6=crossover_type  # "1= 1-Point 2=2-points  3=3-points Uniform Crossover"
7=mutation_type   # "0= fixed   1=variable ==> STAGNATION TYPE"
8=mutation_factor # "1= One gene mutation 2= Two gene mutation 3= Three gene mutation"
9=adaptive_type   # "Adaptive step size [0= None 1= Adaptive]
10=training_size  # "Training set size [0= None 1= Search Space Percentage 2=NEVAL Percentage]"
11=test_size       # "Test set size [0= None 1= Search Space Percentage 2=NEVAL Percentage]"
12=svm_type        # "SVM type [0=none 1= C-SVC 2=nu-SVC 3=one-class SVM 4=epsilon-SVR 5=nu-SVR]"
13=kernel_type    # "SVM Kernel [0=linear 1=polynomial 2=radial basis-RBF 3=sigmoid]"
14=c_parameter    # "Set the SVM Cost parameter C of C-SVC, epsilon-SVR, or nu-SVR (default 1)"
15=gama_parameter # "Set the SVM gamma parameter G in kernel function (0=default 1/k)"
16=epsilon_par    # "Set the epsilon P in loss function of epsilon-SVR (0=default 0.1)"
17=cross_val       # "Cross validation flag [0= none 1= yes (standard)]"
18=agressive_cross# "Aggressive cross validation flag [0=none 1=yes]"
</COMMENTS>
#
<AIDA>
#  "1   2   3   4   5   6   7   8   9  10  11  12  13  14  15   16   17  18"
<KERNEL>
GA-STD - 1 - 1 - 1 - 1 - 1 - 0 - 1 - 0 - 0 - 0 - 0 - 0 - 1 - 0 - 0.1 - 0 - 1
GA-MOD - 1 - 2 - 2 - 4 - 2 - 1 - 2 - 1 - 0 - 0 - 0 - 2 - 10- 0 - 2.0 - 0 - 1
GA-SVM - 3 - 3 - 1 - 4 - 2 - 0 - 2 - 0 - 1 - 1 - 1 - 2 - 50- 0 - 4.0 - 1 - 0
</KERNEL>
</AIDA>
```

Fig. 5.26 Optimization kernel configuration file

"GA-SVM". At least one line should be presented for the correct functioning of GENOM. The command to execute a single optimization in 5 runs and respective simulation result is showed in Fig. 5.27. Each line depicts the run number, *#Run*, the number of evaluations in each run, *#nEvals*, final fitness value, *#Fitness*, simulation time, *#wTIME*, existence of feasible solution, *#FEAS*, and existence of a solution, "*#SOLUTION*", found at generation, "*found_@*". A feasible solution satisfies all designer rules but may miss one performance requirement, on contrary, if a solution is found, all designer rules, as well as, all the performance specs are satisfied.

```
/*************************************************************************/
AGPAR.h - configuration file
Copyright (C) 2005 by Manuel Barros, fmbarros@ipt.pt
/*************************************************************************/

Num_of_runs   5  #number of runs

<AIDA>
#  "1   2   3   4   5   6   7   8   9  10  11  12  13  14  15   16   17   18"
<KERNEL>
GA-MOD - 1 - 3 - 3 - 4 - 2 - 1 - 2 - 1 - 0 - 0 - 0 - 0 - 1 - 0 - 0.1 - 0 - 1
</KERNEL>
</AIDA>
<EOF>
```

RESULTS:

```
                    - PLOT OUTPUT DATA in each run -
---------------------------------------------------------------------------------
#Run  #nEvals    #Fitness   #wTIME      #FEAS  #found_@   #SOLUTION   #found_@
---------------------------------------------------------------------------------
  1       288  7.542e-02   32.87s        Y     10 (gen)          Y    14 (gen)
  2       512  1.162e-01   53.61s        Y     22 (gen)          Y    28 (gen)
  3      1088  7.881e-02  114.81s        Y     56 (gen)          Y    64 (gen)
  4       608  3.428e-02   59.80s        Y     24 (gen)          Y    34 (gen)
  5       640  9.562e-02   61.64s        Y     28 (gen)          Y    36 (gen)
```

Fig. 5.27 A single kernel configuration and results.

5.3 Data Flow Management

In a design automation tool there is a need to handle two types of data structures, one, to manage the circuit's database and the other to manage the simulation data. A good definition of the data structure can lead to efficient data management and improvements in reusability. For instance, the simulation measures, the performance parameters database, the sub-circuits blocks, the testbenches and the technological files are likely to be shared or reused, avoiding the redefinition of circuit's information. In the same way, the data management of simulation data from the synthesis process can also be improved due to the need to control and to establish relations between the huge amount of simulation data, normally, produced from the optimization process, the need to cope with the variety of file formats from different simulators or even a simple access to the simulated data of a specific circuit simulation.

In GENOM, the circuit's database is managed externally by AIDA framework but the management of the simulation data is GENOM's responsibility. When used as a standalone application, GENOM requires the input files illustrated earlier in Fig. 5.6.

The next two sections explain how GENOM manages the data and structures.

5.3.1 *Input Data Specification*

The preferential method to input all the data specification is through a GUI, otherwise the required files have to be manually generated. The GENOM graphical user interface presented in Sect. 5.2.2 inherits some methods of AIDA framework, and, as a result, takes advantage of its technology, namely the data management and data structure used to create and maintain a circuit's library. A multilayered architecture structure organized in tables with relational data, as illustrated in Fig. 5.28 and Fig. 5.29, is used to store the information concerning the circuits introduced through the graphical interface and the data provided by the optimization tool for data visualization. The next screenshots show the input data specification of the filter depicted in Fig. 5.29.

id_circuit	name	categoryId	circuitType	behaviour	bahaviourName
219	TestBench AC Analysis	13	TestBench	22	null
221	Filtro Elíptico de 2ª Ordem	15	Circuit	18	Low Pass
225	stageInfo	-1	stageInfo	(Null)	
226	stageInfo	-1	stageInfo	(Null)	
227	innerFlowInfo	-1	innerFlowInfo	(Null)	

id_designParameters	circuitId	instanceName	name		minValue	value	maxValue		relation	valueString
8239	221		R1		1e3	1e3	1e3		null	null
8240	221		R2	...	1e3	1e3	1e3	...	null	null
8241	221		R3		1e3	1e3	1e3		null	null
8242	221		R4		1e3	1e3	1e3		null	null
8243	221		R5		1e3	1e3	1e3		null	null
8244	221		R6		1e3	1e3	1e3		null	null
8245	221		C1		1e-12	1e-12	1e-12		null	null
8246	221		C2		1e-12	1e-12	1e-12		null	null
8247	221		C3		1e-12	1e-12	1e-12		null	null
8248	221	XA1	RL	...	0.0	0.0	0.0	...	null	R1.R2/(R1+R2)
8249	221	XA1	RC		0.0	0.0	0.0		null	C2
8250	221	XA2	RL		0.0	0.0	0.0		null	R4

Fig. 5.28 The circuit and the parameter tables filled with data from an elliptic filter

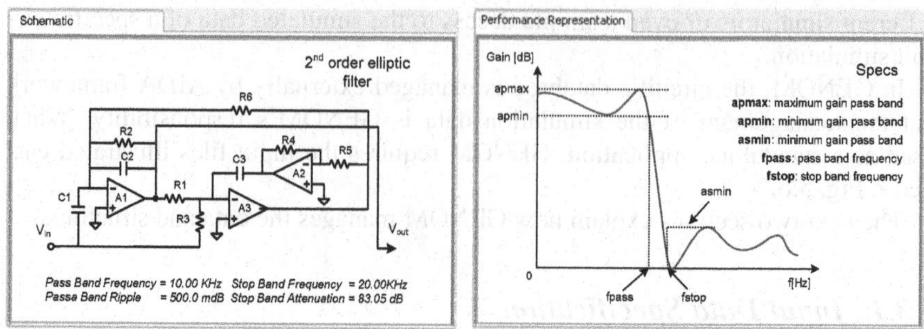

Fig. 5.29 2nd order Elliptic filter section and performance specs.

Essentially, the insertion of a new circuit requires the electrical schematic, a netlist, a technological file, the device parameters, the sub-circuits, the performance parameters and the corresponding measure functions. There are a lot of parameters with different nature associated to a circuit, so all information was arranged (split) in a meaningful storage of well-structured information. The first layer consists in the insertion of elementary data that defines a circuit. The table, at the top of Fig. 5.28, for example, stores the key of circuit identifier (*221*), the name (*Elliptic Filter of 2th Order*), the category (*Filter*, *OpAmp*, etc) of the circuit, the type of circuit (*Circuit*, *testbench*, etc) and the behavior class (*Low pass*). The design parameter table, at the bottom of the Fig. 5.28, represents the parameters table that characterizes each component from the netlist. There is a unique key that identifies each parameter (*8239*) plus the remainder characteristics and it is associated to the circuit where it belongs (*circuited=221*). This table is composed by a long list of parameters which includes a field that marks this component for optimization, another one indicates if the component is matched with any other or not (*matchComponent*) and the correspondent matching value (*matchRelation*), above others.

The Fig. 5.30 shows the relational tables used to store the performance parameter information. The definition of performance parameters which can be measured in a circuit constitutes one critical step in the GENOM development as will be explored in the next section. Meanwhile, it will be explained how performance parameters and function measures are treated in GENOM.

The first step consists in the selection of the desired performance parameters (*apmin*, *apmax*, *asmin* and *stop band frequency*) for the chosen circuit, from the library of available performance parameters (see top table of Fig. 5.30). To avoid duplication of information, these parameters are stored in the table of design parameters composed by a unique identifier (*id_designPerformance*) and the performance parameter identifier (*performanceId*), for example, the key 28 corresponds to pass band maximum gain of the circuit in question (*circuitId*). The next fields are accounted for the definition of the global objectives of the circuit. For example, to specify that the pass band maximum gain of the filter in question should be inferior to 0.5 dB, the introduced values should be defined as the value

id_globalPerformance	name	description	unit
18	DcGain	Dc Gain	dB
19	Phase_Margin	Phase Margin	degree
20	GainBandWidth	Gain badnWidht product	dB
21	Power	Power	
28	Apmax	Pass Band max gain value	dB
29	Apmin	Pass Band min gain value	dB
30	Asmin	Stop Band min gain value	dB
31	fstop	Stop band frequency	Hz
32	fpass	Stop band frequency	Hz

id_designPerformance	performanceId	circuitId	relation	value	currentValue
99	28	221			
101	29	221			
102	30	221			
103	31	221			
104	32	221			

id_measures	circuitId	performanceId	name	definition	analysisType
1	221	28	apmax	max vdb(vout) from= freqIni to= fpass	AC
2	221	29	apmin	min vdb(vout) from=freqIni to= fpass	AC
3	221	28	apmax	max vdb(vout) from=fstop to=freqEnd	AC

Fig. 5.30 Performance parameters and measures functions table

"*maximum*" in field "relation" and the value 0.5 in the field "*value*". The column "*currentValue*" is used to store the value generated by the simulation tool or optimization.

The last step consists in the definition of the measure functions or simply measures which allow the determination of the performance parameter values. In the example considered above three measures for AC analysis are defined. The measures (*id_measures*) are associated to one circuit (*circuitId*) 221 and one performance parameter. They are characterized by a specific name and defined (field "*definition*") in HSPICE format.

5.3.2 Evaluation/Simulation Data Hardware

The quest behind GENOM tool is to provide the designer with an easy access to most relevant simulated data assent in a model of efficiency and precision of results. A block level representation of the simulation data flow in GENOM is exhibited in Fig. 5.31.

The data flow management is explained in three moments of simulation. The first three blocks of Fig. 5.31 cover the setup phase using the circuit management explained in the preceding section.

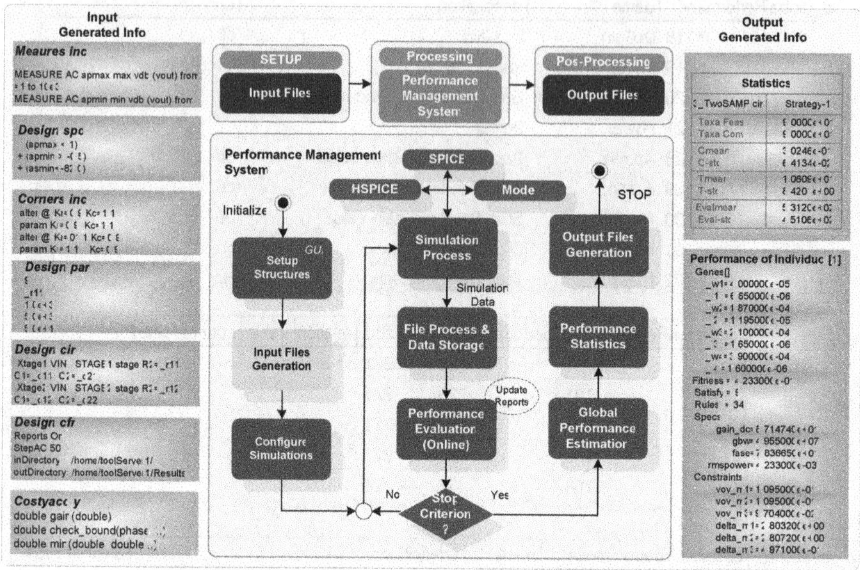

Fig. 5.31 The simulation data management system overview

The second moment is achieved during circuit synthesis process. Here, a parser was created to interpret the language of a circuit specification file and automatically compute the cost function value giving as input the performance parameters of the circuit and the formulations of the cost membership functions. The parser implementation was based in the *Lex* and *Yacc* [4] generation tools so that it is represented by a set of combined grammatical and lexical rules.

The last moment involve the use of built in functions to filter, process and display statistical data from the optimization process either in text or in graphical mode. The primary advantage of text files is that they are very flexible and easy to use. They can be any length, and can accommodate the information to any type of layout and allow the use of database techniques to query a text file.

The principal method of data access involving optimization algorithm and circuit simulator take advantage of the plotting facilities generally found in most electrical simulators. All output variables of interest can be printed in output files using the command ".PRINT" or equivalent. The data format of the response is generally organized in tabular form as depicted in Fig. 5.32. It shows the AC characteristics of the magnitude of voltage and phase in the output node of a filter for a given range of frequencies.

In order to access the data in a file, a file parser is implemented (file process block in Fig. 5.31). The use of file parsing techniques allows the extraction of any necessary information and its employment for later processing. GENOM provides built in functions to view the data in graphic mode version (bode plot characteristics and the cost function evolution). In command mode version, only the

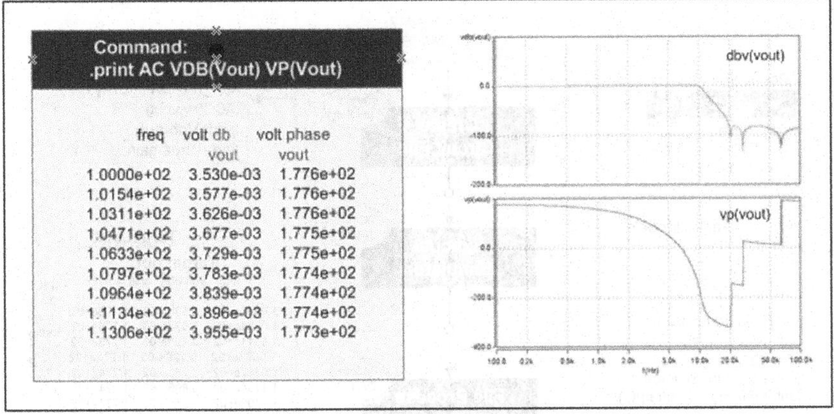

Fig. 5.32 AC analysis in the output node of a filter

extracted plotting files are created, allowing its final treatment with external graphical tools like Avanwaves® [5] or CosmosScope® [5]. The optimization with HSPICE simulator has an extra option that can be automatically invoked to visualize the waveforms in CosmosScope®. The processing of data employing circuit simulators with the purpose of performance estimation employs the same general principle but will be explained next.

5.3.3 Output Data

The entire mode of operation ranging from the moment a chromosome is ready to evaluation until it attains the cost function value will be explained in the following steps and supported by Fig. 5.33.

Step1 - As soon as a new candidate chromosome is submitted to evaluation process, a parser algorithm replaces the optimization parameters values in the target netlist with new ones corresponding to the genes of the chromosome. The "*target.cir*" netlist file is changed.

Step 2 - The new circuit netlist is submitted to electrical simulator (SPICE/HSPICE) producing in the output file (*target.out* or *target.lis*) a long list of simulation data including the matrix of variables and values of interest, and normally the performance parameters resulting from the simulation. This point diverges from simulator to simulator. In SPICE the type of variables are within the scope of command ".PRINT". The HSPICE simulator is more flexible because it incorporates a new command called measures, which gives the user more freedom to print and customize user-defined electrical specifications of a circuit. Actually, this is the preferred method to pass information between HSPICE and GENOM, since in the output file there is only the answer to the requested measures left, thus resulting in a compact file and allowing a more efficient access.

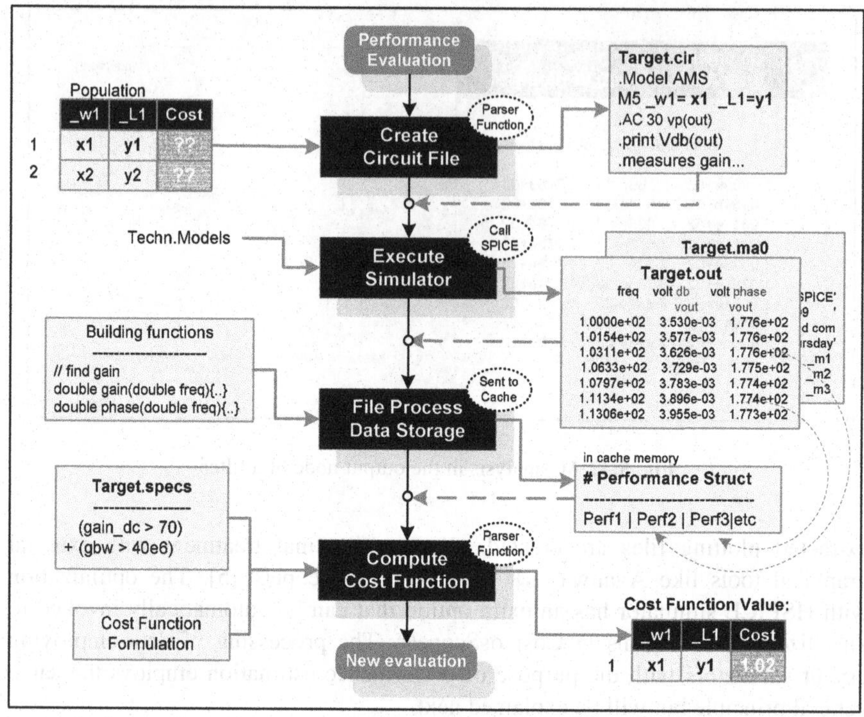

Fig. 5.33 The performance evaluation data flow

Step 3 - Next, a set of built in functions extracts the data information matrix stored in one or more output files and sends it to cache memory structures for fast manipulation. When the required information is not explicit stored, a new built in function is created to compute its value. At the end of this step, all necessary parameters needed to compute the cost function, are organized in memory by the order they appeared in targets specs file.

Step 4 - Finally, the cost function value is automatically computed with the help of a new cost parser function based on the compiler Lex and Yacc (details in sub-Sect. 5.3.3.1). Simultaneously, it collects a set of statistical data that is important to control the optimization algorithm, such as, the number of satisfied solution, the number of violated constraints, the corner's information, etc.

5.3.3.1 The Simulation and Equation Based Cost Function Parser

This section explains the parser implementation behind the cost function computation. The main purpose of the parser is to create a mechanism able to interpret the language of a circuit specification file and automatically compute the cost function value giving as input the performance parameters of the circuit and the formulations of the cost membership functions. The parser implementation was based in the Lex and Yacc generation tools so that it is represented by a set of combined grammatical and lexical rules as illustrated in Fig. 5.34. The Lex description file identifies a series of symbols (logic and arithmetic operators), regular (mathematical functions and built in functions) and transforms them into tokens (reserved word for the language). Once this transformation is done, the YACC syntaxical analyzer interprets this stream of tokens and converts it into a meaningful grammar. With this specification, the GENOM's parser not only is able to interpret more traditional circuit specification files (based on logic and arithmetic operators, see "*target.specs*" in Fig. 5.35) but also specification files based on user defined equations (equation-based). The user defined equations can be expressed through basic mathematical functions ('*Fabs*', '*SIN*', '*SQRT*', '*POW*', etc) or by more sophisticated built in functions such as 'gain()', 'phase()', '*get_Value_Cache()*', *min()*, etc. For example, the function "*double gain(double freq)*", finds the gain corresponding to frequency from the output file of a SPICE simulation. If the performance measures are already in cache memory then the "*get_Value_Cache()*" function can be used instead.

Fig. 5.35 gives a simplified macro view of the actions taken automatically by the parser machine to carry out a single performance parameter. When the "*cost_Calc()*" function triggers the process, the first line of the design specification file is ready for parser analysis. The expression "*(gain_dc>70)*" is evaluated and the identifier "*gain_dc*" must be resolved first. Since "*gain_dc*" expression did not match any of the parser reserved word, it is interpreted as a performance parameter whose value should be read from memory with "*get_Value_Cache()*". Then, the obtained expression '90 > 70' is resolved by executing a set of operations specified by the operator '>'. One of these operations performs a call to the membership functions that translate the impact of this measure in the overall performance. Then the process is repeated line by line until the end of the target specification file.

Lexical Rules "Costlex.l"

```
%option noyywrap
ID [a-zA-Z][_a-zA-Z0-9]*
VAR _[a-zA-Z][_a-zA-Z0-9]*
FPNUM (([0-9]+)|([0-9]*(\.[0-9]+)?([eE][-+]?[0-9]+)?))
%%
"fabs"    { return FABS; }
"sin"     {  return SIN; }
"sqrt"    { return SQRT; }
"pow"     { return POW; }
"gain"    { return GAIN; }
"phase"   { return PHASE; }
"f_gain"  { return F_GAIN; }
"f_phase" {return F_PHASE; }
"check_bound"     { return CHECK_BOUND; }
"verify_bound"    {  return VERIFY_BOUND; }
"min" {return MIN; }
{ID}  { yylval.name = yytext; return ID; }
{VAR} { yylval.name = yytext; return VAR; }
{FPNUM} { yylval.value = atof(yytext); return NUM; }
"+"   { /*printf("+\n");*/ return '+'; }
"-"   { /*printf("-\n");*/ return '-'; }
"*"   { /*printf("*\n");*/ return '*'; }
"/"   { /*printf("/\n");*/ return '/'; }
">"   { /*printf(">\n");*/ return '>'; }
"<"   { /*printf("<\n");*/ return '<'; } ...
```

Grammatical Rules "Costyacc.c"

```
cost: exp { cost_val = $1; }
exp: exp '+' term {
          $$ = $1 + $3; }
    | exp '-' term {
          $$ = $1 - $3; }
// ...
exp: exp '>' term {
              var=
getValue_Cache($1);
  if (var-$3 >= 0.0)
  {  $$=0.0;
    SAT_SPECS++;
              } else {
   $$=fabs(mf_GTA(var,$3));
              $$=getFit($3,var);
} ...
double gain(double);
double phase(double);
double f_gain(double);
double f_phase(double);
double check_bound(double, ..);
double getValue_Cache(int);
double find_indexMeasure(...);
double verify_bound(double, ...
double min(double, double,..); ...
```

Grammatical Rules "Costyacc.c" (cont.)

```
...
func: FABS '(' exp ')' {
      $$ = fabs($3);
   }
  | SIN '(' exp ')' {
          $$ = sin($3);
   }
  | SQRT '(' exp ')' {
          $$ = sqrt($3);
   }
  | POW '(' exp ',' exp ')' {
          $$ = pow($3, $5);
   }
  | GAIN '(' exp ')' {
          $$ = gain($3);
   }
  | PHASE '(' exp ')' {
          $$ = phase($3);
   }
  | F_GAIN '(' exp ')' {
          $$ = f_gain($3);
   }
  | F_PHASE '(' exp ')' {
          $$ = f_phase($3);
   } ...
```

Fig. 5.34 Cost function parser overview

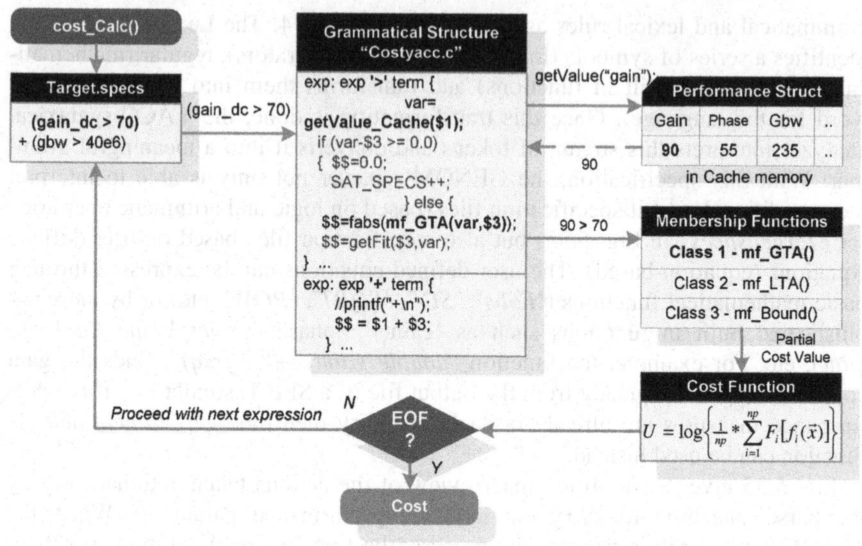

Fig. 5.35 Processing of performance parameters

The parser mechanism allows the implementation of a uniform methodology to access and manipulate data from several sources using simple structures, like the precedence of the operators, their layout, and other grammatical rules which may include built in functions. The use of built in functions allows the integration of new simulators maintaining always a common interface to evaluation of performance parameters.

5.4 Conclusions

This chapter discussed the design architecture, methodology and design implementation of GENOM optimizer tool. The main building blocks included in GENOM are the optimization kernel, the evaluation module and the Graphical User Interface.

The optimization kernel is available with several approaches including the GA standard approach, the modified GA-MOD and the hybrid approach GA-SVM incorporating a learning model based on SVMs.

GENOM was designed to integrate SPICE like simulators, deal with equation based problems and interact with a learning SVM machine. A flexible parser machine was developed to maintain a common interface of the evaluation module allowing the access and manipulation of data from different simulators.

The graphical user interface that controls the inputs and outputs of the system allows the visualization of iterative progress reports. With this feedback an experienced user can assume an active part in the optimization process because he owns some vital information that allows him to twinkle some design parameters during the search mechanism.

References

[1] Barros, M., Neves, G., Horta, N.C.: AIDA: Analog IC design automation based on a fully configurable design hierarchy and flow. In: Proc. 13th IEEE International Conf. on Electronics, Circuits and Systems, pp. 490–493 (2006)
[2] Lourenço, N., Horta, N.C.: LAYGEN – An evolutionary approach to automatic analog IC layout generation. In: Proc. IEEE Conf. on Electronics, Circuits and System, Tunisia (2005)
[3] Lourenço, N., Vianello, M., Guilherme, J., Horta, N.C.: LAYGEN – Automatic layout generation of analog ICs from hierarchical template descriptions. In: Proc. IEEE Ph. D. Research in Microelectronics and Electronics, pp. 213–216 (2006)
[4] Niemann, T.: A compact guide to Lex & Yacc (2004), http://epaperpress.com/lexandyacc/ (Accessed March 2009)
[5] Synopsys Inc, CosmosScope-Waveform analysis (2009), http://www.synopsys.com (Accessed March 2009)

6 Optimization of Analog Circuits and Systems – Applications

Abstract. In the previous chapters there was a description of the optimization methodology and the supporting tool that simplifies the design tasks of analog integrated circuits. The developed design optimization tool, GENOM, based on evolutionary computation techniques and incorporating heuristic knowledge on the automatic control mechanism was combined efficiently with a learning strategy based on SVM to improve the convergence speed of the optimization algorithm. This chapter demonstrates the capabilities and performances of the implemented design optimization methodology when applied to several analog synthesis experiments and provides some insight into factors that affect the synthesis process. Several state of the art circuit blocks will be introduced and optimized for performance and efficiency. Particularly, the performance and effectiveness of GENOM optimizer will be compared with one important reference tool.

6.1 Testing the Performance of Analog Circuits

Operational amplifiers (OpAmps) are the fundamental building blocks of many analog and mixed-signal systems. OpAmps arranged in structures of different levels of complexity are used to realize functions ranging from dc bias generation to high speed amplification or filtering. Table 6.1 presents the general characteristics [1] of some of the OpAmps that will be covered in this chapter.

Table 6.1 General comparison

	Gain	Output Swing	Power Dissipation	Speed	Noise
Two Stage	Medium	Medium	Low	High	Low
Folded Cascode	Medium	Medium	Medium	High	Low
Telescopic	High	Low	Medium	High	Low
Gain-Boosted	High	Medium	High	Medium	Medium

Simulation and testing of CMOS Opamps involve the measure of several performances parameters such as open-loop gain, open-loop frequency response (including phase margin), input-offset voltage, common-mode gain, common-mode rejection ratio (CMRR), power-supply rejection ratio (PSRR), output resistance, noise, output swing, power dissipation and transient response including slew rate.

M.F.M. Barros et al.: Analog Circuits and Systems Optimization, SCI 294, pp. 139–186.
springerlink.com

Special configurations and techniques are necessary to acquire these measurements. The testbench configurations supply the environment (stimulus, load, supplies, etc.) in which the circuit is to be tested. Fig. 6.1 presents the testbench configurations considered for the selected examples [2].

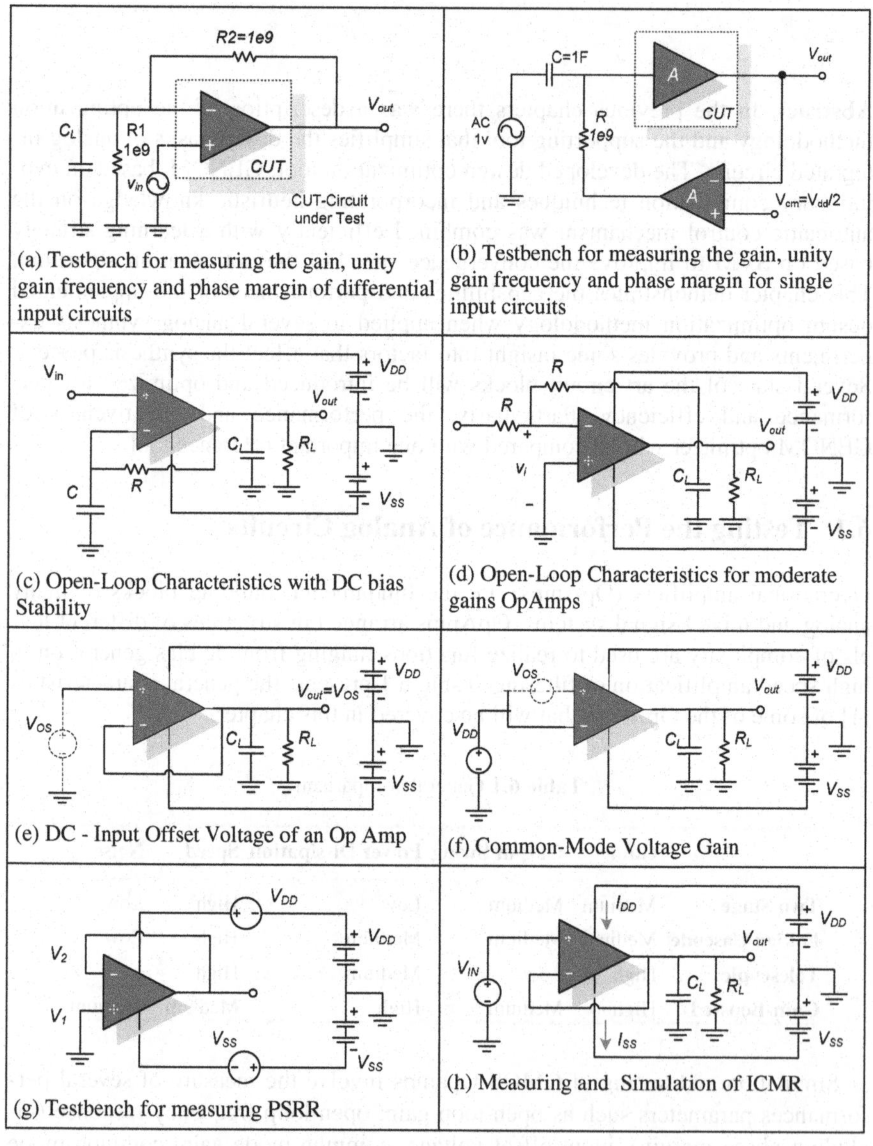

(a) Testbench for measuring the gain, unity gain frequency and phase margin of differential input circuits

(b) Testbench for measuring the gain, unity gain frequency and phase margin for single input circuits

(c) Open-Loop Characteristics with DC bias Stability

(d) Open-Loop Characteristics for moderate gains OpAmps

(e) DC - Input Offset Voltage of an Op Amp

(f) Common-Mode Voltage Gain

(g) Testbench for measuring PSRR

(h) Measuring and Simulation of ICMR

Fig. 6.1 Testbenches to measure the performances values

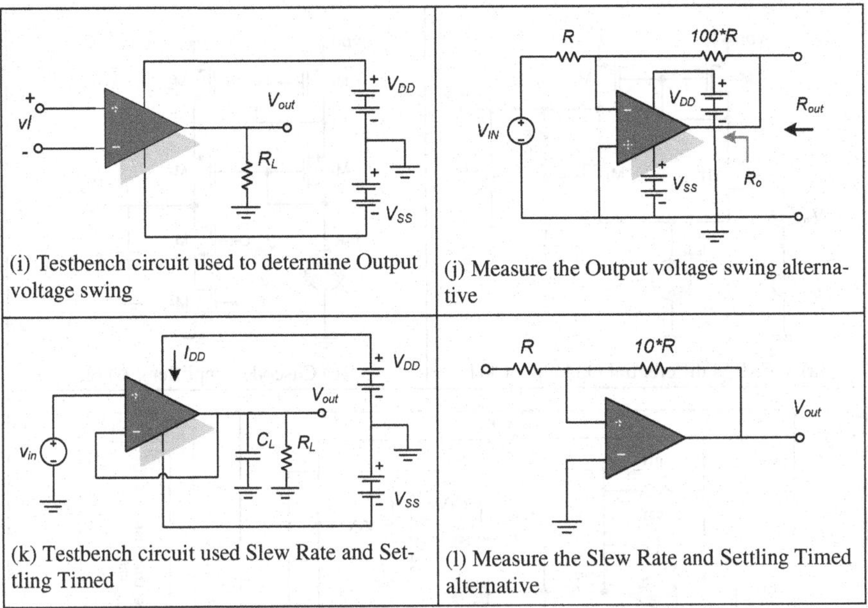

(i) Testbench circuit used to determine Output voltage swing

(j) Measure the Output voltage swing alternative

(k) Testbench circuit used Slew Rate and Settling Timed

(l) Measure the Slew Rate and Settling Timed alternative

Fig. 6.1 *(continued)*

6.2 Testing the GENOM – Selected Circuit Topologies

Since analog benchmark circuits are still unavailable for synthesis purposes, the first testing circuits were collected from the well-known class of CMOS operational amplifiers and also include a low pass elliptic filter listed in Table 6.2 ordered by circuit complexity. OpAmps and filters are fundamental building blocks often employed in analog circuit design applications. Each circuit includes appropriate testbenches to obtain the desired performances parameters measures. The testbench circuit configuration of Fig. 6.2 a), b) and d) were used in these experiments to determine the open loop gain, unity gain frequency, phase margin and power consumption for the single ended circuits. The filter specifications are different and will be defined later. All OpAmp circuits examples were designed using a 0.35-μm AMS (Austria Mikro Systems International AG) CMOS technology process with a supply voltage of 3.3V but the optimization process is fully independent from technology.

The design first step is to determine the design parameters, the functional constraints of the problem and the performance objectives for each topology. Table 6.2 describes the complexity level for each test circuit. In this study, the design parameters are composed by the lengths, widths and/or multiplicity of transistors and are constrained by the ranges in geometry defined in Table 6.3.

Once the parameters have been defined, the GA chromosome can be constructed representing an individual or a candidate solution. The optimization

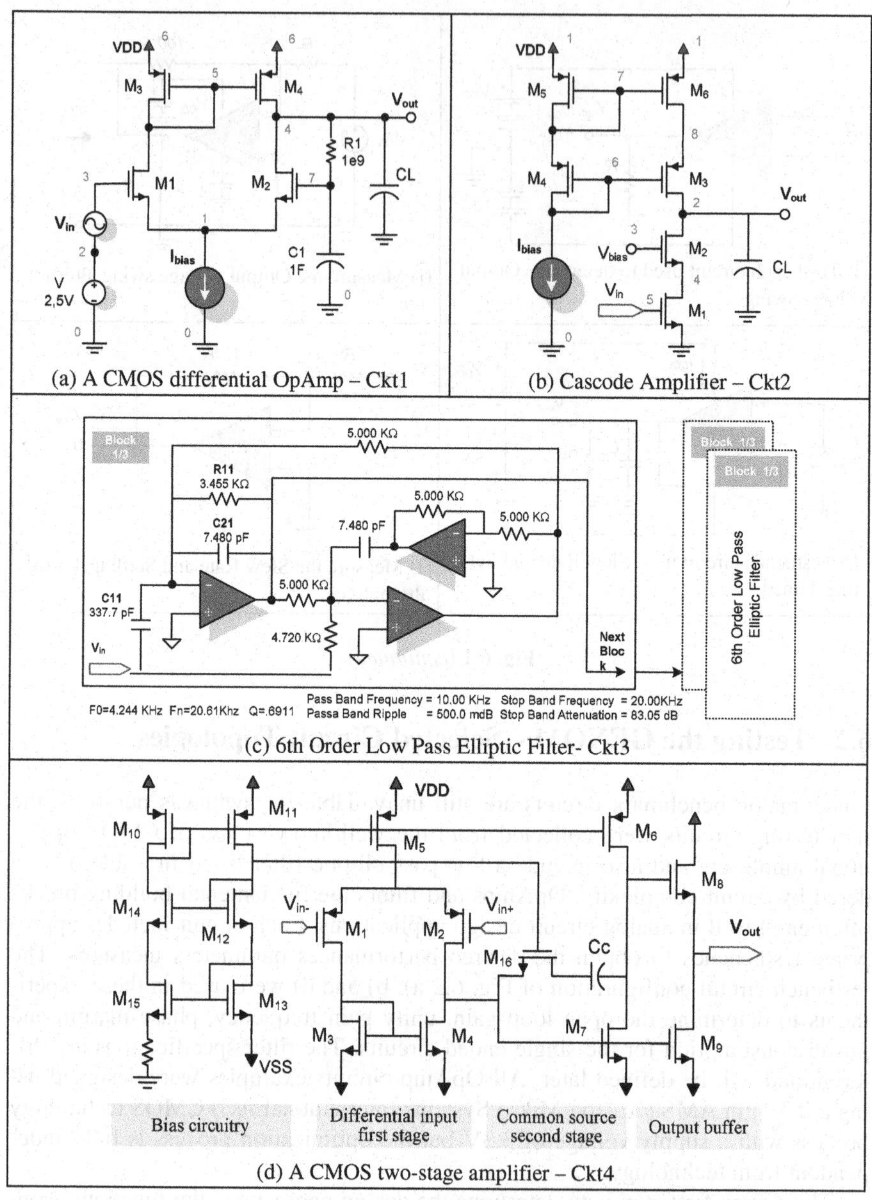

Fig. 6.2 The suite of circuit schematics used in tests

design parameters domain and the adopted technological grid define the complexity of the problem. A set of fundamental designer rules as well as the matching conditions for each design case is depicted in Table 6.4. This set of rules makes up the functional constraints of design optimization. All measures of performance and the conformance level for each designer rules ("*satisfiability*") are obtained by electrical simulation. In each optimization run, the GA generates and optimizes the design parameters according to the fitness function built based on the performance specifications defined in Table 6.5. The total fitness score for each individual was calculated using the fitness function presented in Sect. 3.3.1.

Table 6.2 Class of circuits used in the tests

Ident.	Name	No.Devices	Opt.Var(a)	Constr.(b)	Refs
Ckt1	Differential OpAmp	4	4	12	--
Ckt2	Cascode Amplifier	7	7	12	[3] Exa. 3.11
Ckt3	6th Order Low Pass Elliptic Filter	27	9	--	--
Ckt4	Two-Stage OpAmp	16	10	30	[3] Exa. 5.2

(a) Number of optimization variables (b) Number of constraints

Table 6.3 Design parameters range

Id	W's (μm)[a]	L's (μm)[a]	Ibias (μA)	Search Space
Ckt1	[1, 400,1]	[0.35,10,0.1]	200 fixed	2.137e+10
Ckt2	[1, 400,1]	[0.35,10,0.1]	[10,60,20]	8.905e+14
Ckt4	[1, 400,1]	[0.35,10,0.1]	none	8.220e+18

(a) Note: all parameters ranges means [min, max, grid size] respectively

Table 6.4 Matching and technology constraints details

Id	Techn. Constraints / Matching	$V_{GS} - V_T$ [a] [Min - Max]	$V_{DS} - V_{DSAT}$ [b] Min / Max
Ckt1	M1=M2	[50-200] mV	> 50 mV
	M3=M4	[100-300] mV	> 50 mV
Ckt2	M1=M2;	> 50 mV	> 50 mV
	M3=M4=M5=M6	[100-300] mV	> 50 mV
Ckt4	M1=M2;	> 50 mV	> 50 mV
	M3=M4=M7;	< 200 mV	> 50 mV
	M8=M9;	< 200 mV	> 50 mV
	M10=M11=M12 =M14	< 200 mV	> 50 mV

(a) Overdrive voltages (b) Drain-sources voltages margin

Table 6.5 Specifications/requirements

Id	Gain	GBW	Phase	Power
Ckt1	> 50 dB	>40 MHz	60°<Ph<90°	Min (mW)
Ckt2	> 70 dB	>25 MHz	60°<Ph<90°	Min (mW)
Ckt4	> 65 dB	>20 MHz	60°<Ph<90°	Min (mW)

The optimization algorithms were all initialized with the following default parameters listed in Table 6.6. In particular, the *GA-STD*, *GA-MOD* and *GA-SVM* will be used in the following experiments. GA-STD specifies the configuration of a standard GA, whereas, GA-MOD covers the new GENOM kernel, but, only the modified evolutionary module is considered, and finally, the proposed GA-SVM defines the hybrid method composed by the GA-MOD extended by the SVM learning method.

Table 6.6 Optimization algorithm parameters

Algorithm Setup	GA-STD	GA-MOD	GA-SVM
initPOP	64	64	64
popSize (μ)	initPOP/2	initPOP/2	iintPOP/2
Elite members (λ)	initPOP /8	initPOP /8	initPOP /8
Initial Sampling	Random	Latin Hyper Sampling	Grid
Selection	Random	Tournament w/ Feasibility	Tournament w/ Feasibility
Sort	Min. cost	Priority to Feasibility	Priority to Feasibility
Crossover	1-Point Unif.	2-Point Unif.	2-Point Unif.
Mut.Rate	5 % fixed	5 % Dynamic	5 % Dynamic
Kernel type	GA-STD	GA modified	GA+SVM
Training Set SVM	none	none	2000 Unif Sampling points
Early Stop	yes	yes	yes

Some of the common parameters include the initial population size population size (μ=32), elite size (8), initial mutation rate (5%), a 2-pairs tournament-crossover probability in 50% of μ and a normal distribution method for generating the initial population. The stop criterion was here defined as a maximum number of generations or as soon as it reaches the first solution. For this particularly experiments, the SVM meta-parameters were found for the first time model generation and then fixed (regularization parameter C=4, variance σ=1/n). A cross validation method [4-6] for optimal parameter selection will execute automatically for each model update.

6.3 GENOM Convergence Tests

In this section a set of experiments that tests the convergence and performance of GENOM GA-MOD algorithm will be presented. In particular, a simple testbench OpAmp circuit from Fig. 6.2a) will be used in this study. This circuit has 4 independent variables and was synthesized within a 0.35μm, 3.3V technology. Each variable has a reasonable range and all were initialized by a random sampling methodology.

6.3.1 The Analog IC Design Approach

The GENOM design methodology is ruled by two types of objectives: the goals and the constraints. All design goals and all design constraints must be satisfied in order to obtain a circuit, which fulfills the aims of the application. As soon as, a satisfactory solution is found, the optimizer continues his search for the improvement of each goal, while ensuring that the constraints are still satisfied. During the search, it can happen that a candidate solution may satisfy all performance constraints and goals but may not meet the functional constraints or vice-versa. The space of feasible solutions is given by the candidate solutions that belong simultaneously to the performance and feasibility regions. The computation effort spent to find the solution space will increase as more and more performance constraints, design trade-offs, or even process variation parameters are taken into account when designing robust design circuits. Fig. 6.3 and Table 6.7 show the algorithm performance result for the simple OpAmps for 5 runs executed on an AMD X64 2.8 GHz dual core machine and use HSPICE to simulate the circuit and extract performance parameters.

```
==========================================================================================
                 OUTPUT STREAM OF SIMULATION DATA FOR CONFERENCE PROCEEDINGS
==========================================================================================
                              - PLOT OUTPUT DATA in each run -
------------------------------------------------------------------------------------------
#Run  #nEvals    #Fitness  #wTIME    #Perf #Found@  #FEAS  #found_@  #STATUS  #found_@
------------------------------------------------------------------------------------------
 1       128   1.065e-02    8.92s    Y     2 (gen)  Y      4 (gen)   Y        4 (gen)
 2        64   1.062e-02    4.29s    Y     0 (gen)  Y      1 (gen)   Y        1 (gen)
 3        64   1.070e-02    4.84s    Y     0 (gen)  Y      0 (gen)   Y        0 (gen)
 4       192   1.082e-02   13.47s    Y     2 (gen)  Y      2 (gen)   Y        8 (gen)
 5        68   1.079e-02    4.22s    Y     1 (gen)  Y      1 (gen)   Y        1 (gen)
------------------------------------------------------------------------------------------
```

Fig. 6.3 Print screen with statistical data from nominal optimization

Each line from Fig. 6.3 depicts the run number, *#Run*, the number of evaluations in each run, "*#nEvals*", the final fitness value, "*#Fitness*", simulation time, "*#wTIME*", then its followed by three binary values indicating whether a solution satisfies all performance constraints "*#PERF*", all feasibility (designer rules) constraints "*#FEAS*" or both, meaning that a solution was found *#STATUS*=Y at generation "*found_@*".

The "*Perf. Specs*" columns in Table 6.7 mean the fitness, time and evaluation number when the circuit meets all design specs of the problem. In the same way, the "*Specs&Rules*" column represents the same features when the circuit meets all design specs, as well as, and all functional constraints of the problem, considering the nominal optimization with typical working conditions. The "*Corners*" column also represents the same features, in case the circuit meets all design specs and all functional constraints in all corner points of the problem.

Table 6.7 Overall performance measures

	#Fitness			#TIME (s)			#nEVALs		
RUNs	Perf. Specs	Specs & Rules	Corners	Perf. Specs	Specs & Rules	Corners	Perf. Specs	Specs & Rules	Corners
Run-1	1.065e-02	1.065e-02	1.059e-02	6.75	8.92	174.98	96	128	2578
Run-2	1.062e-02	1.062e-02	1.062e-02	4.29	4.29	184.08	<64	<64	2720
Run-3	1.07e-02	1.07e-02	1.061e-02	4.84	4.84	232.93	<64	<64	3440
Run-4	1.073e-02	1.082e-02	1.059e-02	8.10	13.47	260.70	112	192	3888
Run-5	1.079e-02	1.077e-02	1.057e-02	4.22	5.28	232.25	<64	68	3424

The optimization process considering only typical conditions solved the problem quickly, and spent only a very few generations (from 0 to 8) as seen in Fig. 6.3 to achieve the performance specs satisfying all design constraints (rules). However, in corner optimization the number of generations increases for around 15-20 generations. Since each candidate solution for corner analysis requires 9 SPICE simulations (one simulation for each corner point), a minimum of 2578 and a maximum of 3888 HSPICE simulations were performed taking into account all runs.

6.3.2 Testing the Selection Approach

Considering the search space subdivision in performance and feasibility spaces, this experiment tries to answer the question of which selection approach is more efficient to handle analog circuit candidates towards the optimum space. When two candidate solutions are compared, which one is more efficient, the one satisfying all performance specs less 50% of constraints or the one satisfying all design constraints less the 50% of specs? It will be seen in the following experiments the influence of the selection operator materialized in GENOM by the variation of the sort algorithm and the tournament selection scheme.

The following results, depicted in Table 6.8 and Table 6.9, present the effectiveness of the selection operator variants implemented in GENOM optimizer, using the same circuit of Fig. 6.2a) for the corner optimization case. In particular two variants will be tested. The first variant promotes the solutions close to the performance space, i.e., in the pathway to the solution space, and its first goal is to reach the performance space and then move towards the feasibility space (results in Table 6.8). A second variant uses the opposite strategy, the first approach is to reach the feasibility space and after that the performance space (Table 6.8). The performance of these two approaches will be compared with standard approach (Table 6.10).

Table 6.8 Output results for each run - Priority to the performance space

#Run	#nEvals	#Fitness	#wTIME	#PERF	#found_@	#STATUS	#found_@
1	2578	1.060e-02	117.07s	Y	1 (gen)	Y	14 (gen)
2	2720	1.061e-02	130.58s	Y	1 (gen)	Y	15 (gen)
3	3440	1.061e-02	169.89s	Y	3 (gen)	Y	20 (gen)
4	3888	1.060e-02	205.95s	Y	3 (gen)	Y	23 (gen)
5	4144	1.060e-02	199.57s	Y	4 (gen)	Y	25 (gen)

Table 6.9 Output results for each run – Priority to the feasibility space

#Run	#nEvals	#Fitness	#wTIME	#FEAS	#found_@	#STATUS	#found_@
1	6472	1.060e-02	296.48s	Y	1 (gen)	Y	41 (gen)
2	6314	1.059e-02	311.79s	Y	1 (gen)	Y	40 (gen)
3	3024	1.060e-02	158.91s	Y	1 (gen)	Y	17 (gen)
4	4464	1.060e-02	220.19s	Y	1 (gen)	Y	27 (gen)
5	1432	1.060e-02	73.83s	Y	1 (gen)	Y	6 (gen)

Table 6.10 Output results for each run – Standard approach

#Run	#nEvals	#Fitness	#wTIME	#FEAS	#found_@	#STATUS	#found_@
1	3448	5.326e-02	190.46s	Y	1 (gen)	Y	20 (gen)
2	6608	5.317e-02	318.82s	Y	1 (gen)	Y	42 (gen)
3	2160	5.327e-02	125.24s	Y	1 (gen)	Y	11 (gen)
4	2736	5.321e-02	121.55s	Y	1 (gen)	Y	15 (gen)
5	2008	5.322e-02	89.36s	Y	1 (gen)	Y	10 (gen)

In the standard approach, the best-ranked individual will always be the one with the lowest constraints and specs violation in each generation. From the analysis of these results it is verified that the standard ranking approach and the ranking strategy that gives priority to the solutions satisfying performances spaces produces the better results in terms of number of generations or computation time. In average, both strategies have similar performances (e.g., the average number of generations is 19.4 and 17.8 respectively), although the standard approach presents worse variances from run to run (13.1 against and 4.8 for the other strategy). For simple circuits like the one used in these experiments there is no apparent benefit in these two approaches.

However, for more complex circuits the great variance of standard approach will be amplified and will produce undesirable results, as shown in Table 6.11 and Table 6.12 for the fully differential OpAmp with 21 optimization variables and 43 constraints defined in Sect. 6.5.1.

Table 6.11 Output results for each run – Priority to the performance space

#Run	#nEvals	#Fitness	#wTIME	#FEAS	#found_@	#STATUS	#found_@
1	288	7.542e-02	16.60s	Y	7 (gen)	Y	14 (gen)
2	512	1.162e-01	26.10s	Y	9 (gen)	Y	28 (gen)
3	1088	7.881e-02	54.80s	Y	21 (gen)	Y	64 (gen)
4	608	3.428e-02	31.99s	Y	7 (gen)	Y	34 (gen)
5	640	9.562e-02	33.91s	Y	10 (gen)	Y	36 (gen)
6	1920	7.108e-02	93.19s	Y	12 (gen)	Y	116 (gen)
7	640	9.099e-02	51.76s	Y	18 (gen)	Y	36 (gen)
8	832	5.907e-02	64.20s	Y	9 (gen)	Y	48 (gen)
9	1168	3.139e-02	65.11s	Y	20 (gen)	Y	69 (gen)
10	832	1.230e-01	41.25s	Y	7 (gen)	Y	48 (gen)

Table 6.12 Output results for each run – Standard approach

#Run	#nEvals	#Fitness	#wTIME	#FEAS	#found_@	#STATUS	#found_@
1	368	6.846e-02	19.85s	Y	11 (gen)	Y	19 (gen)
2	1328	3.895e-02	64.95s	Y	10 (gen)	Y	79 (gen)
3	448	9.689e-02	23.30s	Y	14 (gen)	Y	24 (gen)
4	1616	3.544e-02	78.14s	Y	17 (gen)	Y	97 (gen)
5	384	1.141e-01	20.08s	Y	14 (gen)	Y	20 (gen)
6	880	1.121e-01	44.30s	Y	22 (gen)	Y	51 (gen)
7	2464	2.413e+00	120.51s	Y	9 (gen)	N	>150 (gen)
8	2464	9.955e-01	142.98s	Y	7 (gen)	N	>150 (gen)
9	528	5.027e-02	27.67s	Y	18 (gen)	Y	29 (gen)
10	2464	1.443e+01	169.66s	Y	14 (gen)	N	>150 (gen)

In several runs, the standard ranking approach is not capable of finding a solution during the specified number of generations (150 in this case) for this nominal optimization problem. The ranking strategy with priority to performance space is able to find a solution in all cases (as noticed in Table 6.11) and, in general, it is more efficient to find a solution in each run.

6.4 Comparing GA-STD, GA-MOD and GA-SVM Performance

The objective of these experiments is to compare the performance of the proposed learning method GA-SVM against the earlier evolutionary approach GA-MOD, as well as, the standard GA-STD. The following case studies do not include the search space decomposition feature and the parallelism in the results analysis.

For all the following examples, the industry HSPICE simulator will be used as the evaluation engine, every time an electrical simulation is required. The testbench circuit configuration of Fig. 6.2 b), c) and d) were used in these experiments following the specifications, constraints and models already defined in Sect. 6.2.

In order to create an accurate SVM Feasibility model the optimization parameter space was uniformly sampled with 2000 points to produce the training set, 20% were used to balance the model class samples and 10% more to the validation data set. The class balance pre-processing module was achieved in two steps. First, by filtering those solutions that belong to regions of the design space that are far from fulfill the technological constraints (undersampling the majority class). Then build a two class feasibility model considering those samples which are close the feasibility region and the samples that really belong to the feasibility region. Next, use it to oversample the feasibility region (increasing the minority class) as well as its frontier as explained in Sect. 4.2.5. After that, a final accurate feasibility model is built to be use in the optimization process.

6.4.1 GA-STD versus GA-SVM Performance – Filter Case Study

The filter circuit shown in Fig. 6.2 c) was optimized according to the performance specifications of Table 6.13. The nine design parameters range and the achieved results concerning device sizes are presented in Table 6.14 using the HSPICE simulator as the evaluation engine.

Table 6.13 Performance specifications/requirements

SPECs	Initial	GA-STD	GA-SVM	Units
Maximum P-Band Ripple	< 1	9.13e-01	7.20e-01	dB
Minimum P-Band Ripple	> -0.5	-1.89e-01	-3.93e-01	dB
Stop Band Attenuation	< -82	-8.25e+01	-8.30e+01	dB

Table 6.14 Design parameter specifications (GA-SVM)

Optimization Parameters	Limits	Results
R11 (Ω) in block 1	[1.0e+3, 5.0e+3]	3.70e+03
C11 (F) in block 1	[250.0e-12, 400.0e-12]	3.15e-10
C21 (F) in block 1	[1.0e-9, 10.0e-9]	8.00e-09
R12 (Ω) in block 2	[7.0e+3, 15.0e+3]	1.13e+04
C12 (F) in block 2	[250.0e-12, 400.0e-12]	3.45e-10
C22 (F) in block 2	[1.0e-9, 5.0e-9]	3.90e-09
R13 (Ω) in block 3	[30.0e+3, 40.0e+3]	3.93e+04
C13 (F) in block 3	[50.0e-12, 100.0e-12]	7.40e-11
C23 (F) in block 3	[1.0e-9, 10.0e-9]	3.10e-09

Table 6.15 Runtime info

	GA-STD	GA-SVM
Optimization Variables	9	9
Number of evaluations to get first solution	1670	1272
Time elapse to get **1st** solution	75s	64s

* In a dual processor core AMD at 2400 MHz running Linux OS.

The obtained performance specs obtained by the GA-STD and GA-SVM methods are included in Table 6.13. Finally, the overall computational times are presented in Table 6.15 and the first solution is the one which satisfies all the performance specs.

Both models GA-SVM and GA-STD obtain feasible solutions as outlined in Fig. 6.4, but with slight differences in time efficiency, about 15-20% of efficiency favorable to GA-SVM, as indicated in Table 6.15. With this optimization methodology the GA algorithm may lose some diversity, however the model will improve dynamically one step after the other, as it can be observed in Fig. 6.5, exploring very well, say aggressively, the performance space.

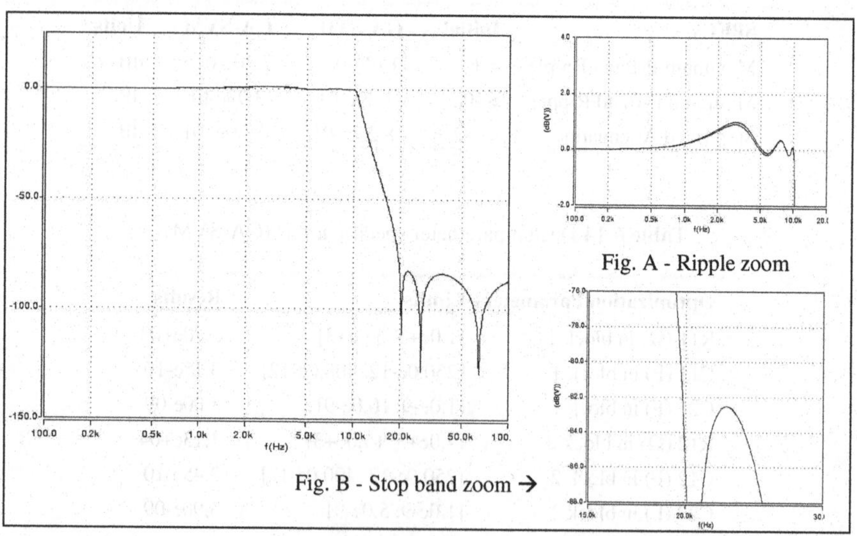

Fig. 6.4 Final Bode plot

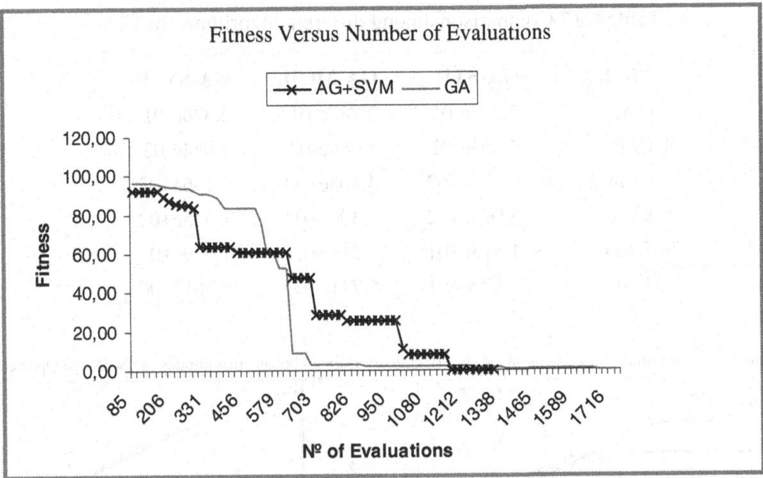

Fig. 6.5 Performance: GA-STD versus GA-SVM kernel

6.4.2 Static GA-SVM Performance - OpAmp Case Study

In this experiment the Ckt2 and Ckt4 OpAmp circuits shown in Fig. 6.2b) and d) were optimized according to the performance specifications of Table 6.5. All statistics measures presented in Table 6.16 and Table 6.17 are the mean and standard deviation obtained over 20 runs. "Cmean" and "Cstd" stand for the mean and the standard deviation of the cost function; "EVmean" and "EVstd" stand for the mean and the standard deviation of the number of evaluations necessary to get the first solution, and finally, the "Tmean" and "Tstd" represent the mean and the standard deviation of the time spent in the optimization process, not included the setup time to build the model in the case of the GA-SVM algorithm.

Table 6.16 Comparison among different algorithms for Ckt2

Cir-1	GA-STD	GA-MOD	GA-SVM
Cmean	9.090e-02	7.476e-02	7.181e-02
Cstd	2.128e-02	6.940e-03	9.646e-03
EVmean	1.888e+02	1.502e+02	7.285e+01
EVstd	8.490e+01	7.043e+01	2.377e+01
Tmean	2.026e+00	1.669e+00	7.275e-01
Tstd	1.109e+00	7.801e-01	3.246e-01

Fig. 6.6 and Fig. 6.7 show the electrical characteristics of the final population and some of the output reports from the optimization tool, respectively.

Table 6.17 Comparison among different algorithms for Ckt4

Cir-2	GA-STD	GA-MOD	GA-SVM
Cmean	2.772e-01	2.787e-01	2.376e-01
Cstd	7.693e-02	5.066e-02	5.034e-02
EVmean	7.216e+02	3.863e+02	4.196e+02
EVstd	3.008e+02	1.300e+02	1.325e+02
Tmean	1.813e+01	1.216e+01	1.029e+01
Tstd	1.179e+01	5.771e+00	4.161e+00

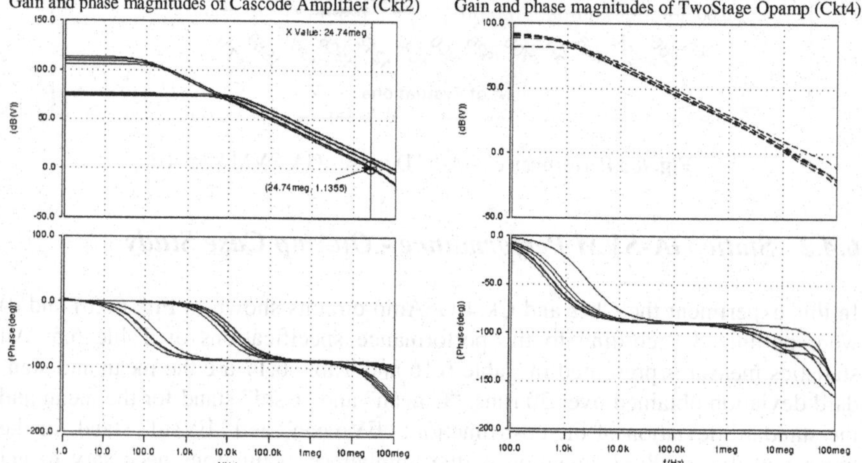

Fig. 6.6 Electrical characteristics from final population

```
- Comparison Among Diferent Algorithms -

|3_TwoSAMP.cir |  Strategy-1 |  Strategy-2 |  Strategy-3 |  Strategy-4 |

|  Taxa Feas   | 1.000000e+02
|  Taxa Conv   | 1.000000e+02

|  Cmean       | 2.772900e-01
|  C-std       | 7.693439e-02

|  Tmean       | 1.813700e+01
|  T-std       | 1.179063e+01

|  Evalmean    | 7.216000e+02
|  Eval-std    | 3.008638e+02

|  Rank        | 1.000000e+00
```

```
        OUTPUT STREAM OF SIMULATION DATA FOR GLVLSI CONFERENCE

                 - PLOT OUTPUT DATA in each run -

#Run  #nEvals  #Fitness   #wTIME   #FEAS  #found_@   #STATUS  #found_@
  1     1232   3.306e-01   28.07s    Y    73 (gen)      Y     73 (gen)
  2      144   2.490e-01    3.49s    Y     2 (gen)      Y      5 (gen)
  3      656   1.889e-01   14.73s    Y     2 (gen)      Y     37 (gen)
  4      656   2.634e-01   17.62s    Y     8 (gen)      Y     37 (gen)
  5      528   2.625e-01   12.12s    Y     1 (gen)      Y     29 (gen)
  6     1312   3.768e-01   30.88s    Y    78 (gen)      Y     78 (gen)
  7      656   2.187e-01   20.30s    Y     5 (gen)      Y     37 (gen)
  8     1152   2.710e-01   37.66s    Y    24 (gen)      Y     68 (gen)
  9     1312   2.445e-01   56.63s    Y    25 (gen)      Y     78 (gen)
 10      800   4.803e-01   16.46s    Y    46 (gen)      Y     46 (gen)
 11      736   1.878e-01   16.66s    Y    13 (gen)      Y     42 (gen)
 12      528   1.796e-01   10.84s    Y     1 (gen)      Y     29 (gen)
```

Fig. 6.7 Output reports from optimization tool (Ckt4)

6.4.2.1 Evaluation Metric

The experiments were executed on AMD X64 2.8 GHz dual core machine and used HSPICE to simulate the circuit and extract performance parameters and the public domain LIBSVM tool [7] as the learning engine. Each algorithm was executed 20 times to acquire the mean and the standard deviation for the evaluation performance. The convergence behavior for the "Two-Stage" OpAmp experiment in one run is presented as an example in Fig. 6.8.

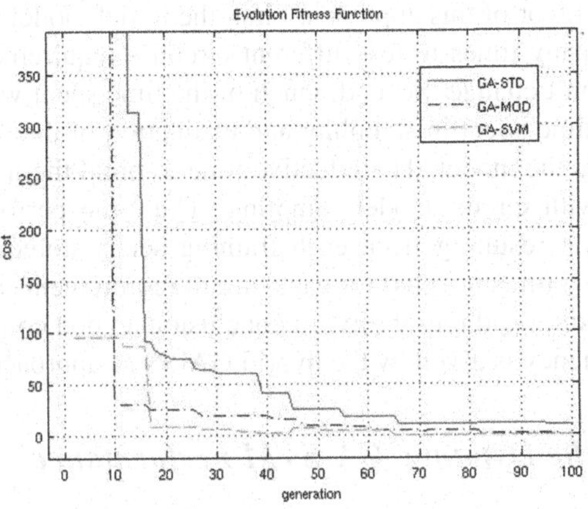

Fig. 6.8 Evolution of the cost function

Analyzing this Fig. and the experimental data displayed in Table 6.16 and Table 6.17 and Fig. 6.9, it is noticeable the good accuracy and lower variance obtained by the GA-MOD and GA-SVM algorithms.

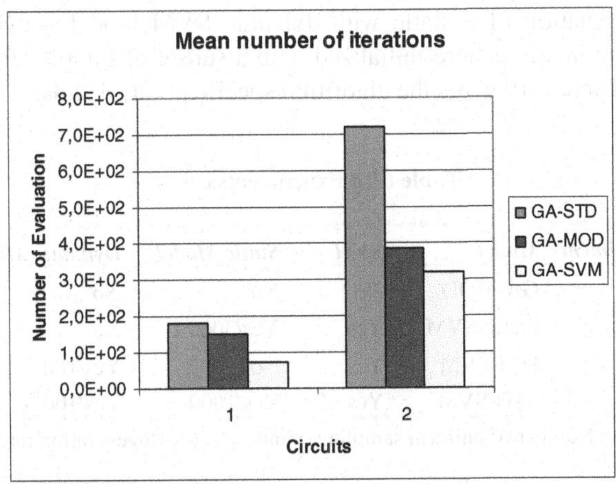

Fig. 6.9 Comparative graph for the required number of evaluations

However, the GA-SVM achieved better results in all cases. Although the GA-STD uses electrical simulation too, the convergence is slower than the others. The algorithms using models are clearly more time efficient if it is not included the algorithm setup time to build the models. Among all the approaches under test, GA-SVM can achieve the lowest cost and the smaller amount of computation time followed by the GA-MOD.

The setup time to build the model, 100 seconds approx. in each of the presented cases, can be problematic at first sight. The means and variances for the GA-SVM would be very different if they were included in statistics. However some points can be clarified in favor of this approach. First, the initial model is build only once and can be used many times to test different circuit's requirements since the parameters ranges don't change. Second, much of the time spent with model generation is due to the time spent in sampling and evaluation of the selected points for training and testing the model. The effective time to build the model is negligible when compared with circuit model sampling. Thus, the performance and constrained information resulting from each training set is stored, it is possible to build a model at any time, adapted for each circuit requirements and allowing posterior model upgrading and reusability. In conclusion a good compromise between accuracy and efficiency is given by the hybrid GA-SVM approach.

6.4.3 Testing the Dynamic GA-SVM Performance

The objective of these experiments is to study the impact of the proposed dynamic SVM model in the optimization process. Our purpose is to compare the performance of several dynamic learning strategies and compare the GA-SVM against the static GA-SVM defined in the previous section, as well as, GA-MOD. Specifically, four experiments defined in Table 6.18 will be performed considering, respectively, the SVM model built before the start of the evolutionary process – static model, SVM model built dynamically, i.e., during evolutionary process, and finally a combination of a static with dynamic SVM model – dynamic model, where the static model is here initialized with a subset of samples from the single static model. Table 6.19 gives the algorithm specifications details.

Table 6.18 Experiments cases

Experiment	*Model*	*SVM*	*Static Model*	*Dynamic Model*
Exp-1	GA-MOD	No	No	No
Exp-2	Static-SVM	Yes	Yes/3000[(a)]	No
Exp-3	Dyn-SVM	Yes	No	Yes/100
Exp-4	S+D-SVM	Yes	Yes/1000	Yes/100[(b)]

(a) Number of uniform sampling points (b) Regeneration rate

Table 6.19 Algorithm specifications under test

Algorithm	*GA-MOD*	*Static-SVM*	*Dynamic-SVM*
Selection	2-Tournament	2-Tournament	2-Tournament
Crossover	2-Point Unif	2-Point Unif	2-Point Unif
Mutation Rate	5% Dynamic	5% Dynamic	5% Dynamic
Kernel type	GA modified	SVM-RBF	SVM-RBF
Training Set	None	3000 Unif Sampling points	None

These experiments use exclusively the two-stage (Ckt4) ampop illustrated in Fig. 6.2, updated with appropriate test benches to allow the measures of the desired performances parameters. All experiments used the same computation resources, specifications and constraints as earlier and also used the same number of runs to extract the mean and the standard deviation for the evaluation performance. The convergence behavior for the two-stage OpAmp experiment in one run is presented as an example in Fig. 6.10.

From the experimental data, displayed from Table 6.20 and Fig. 6.10, it is clear the good accuracy and time efficiency obtained with strategies embedded with SVM models built in offline mode. However, the overhead time to build the static model can be problematic for more complex circuits. Here, the static algorithm takes about 90 seconds approx. to evaluate 3000 uniform samples but in more complex circuits, this number rises considerably. The means and variances to setup the models using static modeling were not included in the final statistics given at Table 6.20.

Table 6.20 Comparison among different algorithms

Algorit.	*Cmean*	*Cstd*	*EVmean*	*EVstd*	*Tmean*	*Tstd*
Exp-1	2.55e-01	4.47e-02	3.95e+02	1.07e+02	1.14e+01	4.04e+00
Exp-2	2.61e-01	4.93e-02	1.48e+02	1.08e+02	3.91e+00	2.28e+00
Exp-3	2.19e-01	4.74e-02	2.74e+02	2.24e+02	6.93e+00	6.58e+00
Exp-4	2.19e-01	5.29e-02	6.22e+02	1.73e+02	1.42e+01	4.31e+00

A different strategy has been taken towards a dynamic building model with data gathered during the early generations. Some configurations were tested as shown in the Fig. 6.10 (b).

This approach can be very sensitive to the value of the regeneration rate value. Using a lower value for the regeneration rate, e.g., 200, originates long processing times because it takes more training samples however a better accuracy model is obtained. A higher sampling rate at early generations causes better convergence but with a slightly increase in execution times. An automatic and dynamic control of the regeneration rate can be added using the information of the quality of SVM model. A good compromise between these two approaches is given by the test case joining the static and dynamic training model behavior (S+D-SVM).

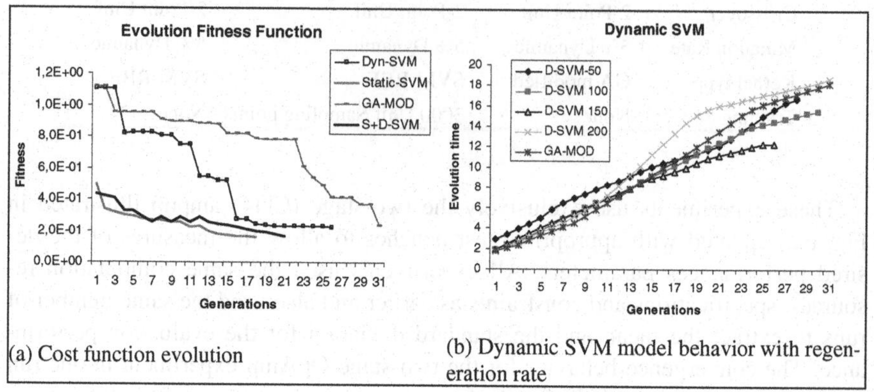

(a) Cost function evolution

(b) Dynamic SVM model behavior with regeneration rate

Fig. 6.10 Comparative performance analysis

6.4.4 *Final Comments*

The proposed approach is a step forward when compared with the simple GA kernel, as it now incorporates performance modeling facilities, allowing an effective pruning of the candidate solutions before being submitted to the heavy time-consumption task of electrical evaluation. The achieved results show significant gains in efficiency and this approach also allows the reuse of the model generated during one optimization process in subsequent optimizations, which is again another significant advantage when compared with traditional approaches, especially in the areas of architecture exploration and synthesis of complex analog blocks.

6.5 General Purpose Circuits or High Performance Circuits Design

In this section, a case study for several high performance circuit designs will be presented passing by the following phases: full schematic, design specifications and constraints, variable ranges, optimization results such as variables size and achieved performance and time statistics. This set of circuits shows GENOM's ability to design high-performance and novel circuit topologies. The design complexity decomposition was optionally not taken into account because the primarily objective is to test the algorithm not the design process.

6.5.1 Fully Differential OpAmp

Fig. 6.11 illustrates the differential amplifier schematics considered to evaluate the performance of the presented optimization technique.

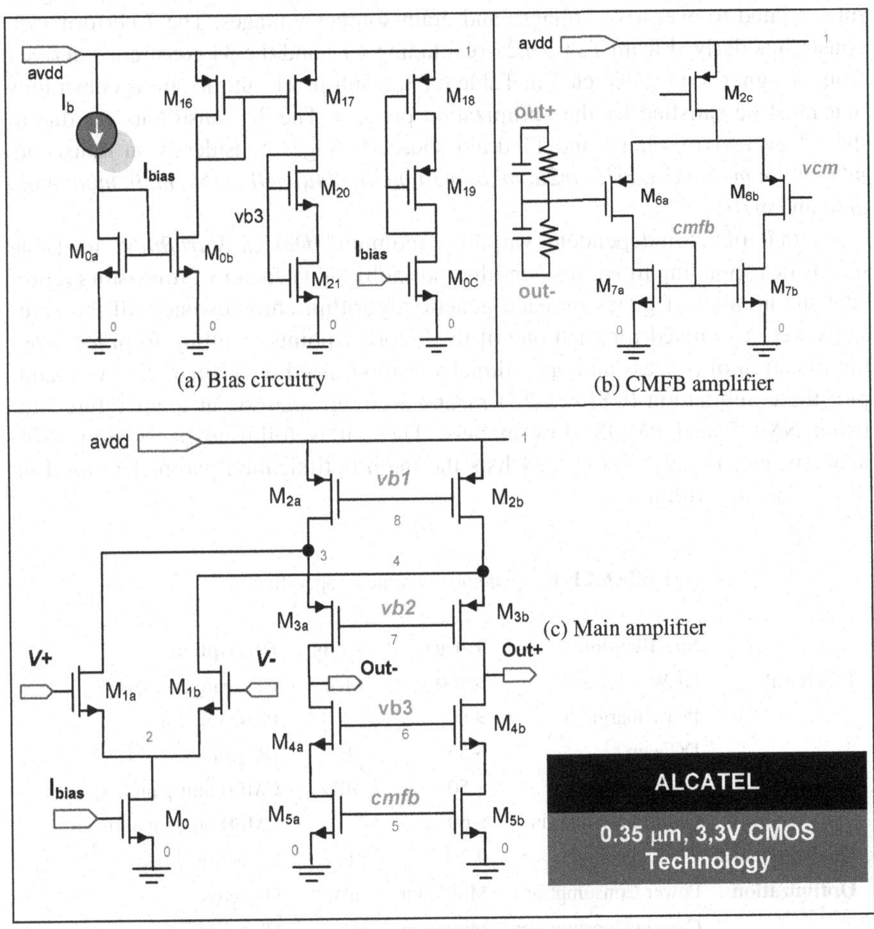

Fig. 6.11 Differential amplifier schematic

The topology, defining the connectivity of device-level components, consists of 25 transistors devices grouped in 3 main functional blocks: the main amplifier with differential input and output, the bias circuit and the common mode feedback circuitry. By looking at the circuit schematic, some groups of transistors like M3a and M3b, for instance, must be matched. Some dependent relations like, the multiplicity factor, m7, of transistors M7a and M7b is equal to m5/2 (this implies that m5 must be pair) must also be verified.

6.5.1.1 Performance Specifications, Input Variables Ranges and Design Space Size

The main objective was to synthesize the presented differential amplifier using the ALCATEL 0.35μm, 3.3V CMOS technology according to the performance specifications, listed in Table 6.21, and always respecting the fundamental designer rules related to overdrive voltages and drain-sources voltages. The 7 performance constraints derived from Table 6.21 (excluding *CL*) and the 34 constraints derived from designer's rules depicted in Table 6.23, result in 41 optimizations constraints that must be satisfied by the optimization process. The 34 constraints are due to the 17 overdrive voltage and 17 drain source voltages considered on transistors *m0a*, *m0b*, *m16*, *m1a*, *m1b*, *m2a*, *m2b*, *m3a*, *m3b*, *m4a*, *m4b*, *m5a*, *m5b*, *m6a*, *m6b*, *m7a* and *m7b*.

A total of 21 independent variables (column "*Design Variables*" in Table 6.23) corresponding to widths, lengths and multiplicity factor of transistors represent the number of genes on each genetic algorithm chromosome. All the solutions were examined for each one of the 9 corner points resulting from the cross combination of process and operational variation listed on Table 6.22. For example, the combination (CSlow,-40º) means a circuit analysis at temperature -40º using NMOS and PMOS slow models. Then, it is followed by (CSlow,+50º) analysis, etc. Finally, Table 6.24 lists the main optimization parameters used on the genetic algorithm.

Table 6.21 Performance parameter specifications

	Specifications	Target	Units	Description
Electrical	GBW	> 100	MHz	Unit-gain frequency
	Phase margin	> 60	º	Phase margin
	DC gain	> 55	dB	DC gain
	CMF GBW	> 50	MHz	CMFB unit-gain frequency
	CMF Phase margin	> 60	º	CMFB phase margin
Environmental	CL (fixed value)	0.2	pF	Capacitive Load
Optimization	Power Consumption	Minimum	mW	Objective
	Current Consumption	Minimum	μA	Objective

Table 6.22 Corners analysis data

Conditions	Variation points		
MOS worst case parameters	CSlow	CTyp	CFast
Temperature Range (º C)	-40º C	+50º C	+120ºC

Table 6.23 Matching and constraints details

Matching				**Constraints**		
Dependent Variable	Design Variable	Range [Min;Max;Step]	Unit	V_{GS} - V_{T} [(a)] [Min - Max]	V_{DS} - V_{DSAT} [(b)] Min / Max	Unit
M0 (_w00, _l00, m02)	_w00	[1; 20; 1]	µm			
M0a=M0b=M0c (_w02, _l00, 1)	_w01	[1; 20; 1]	µm	[100 - 300]	>100	mV
M1 (_w02, _l06, m06)	_w02	[1; 20; 1]	µm			
M2a = M2b=M2c (_w02, _l02, m02)	_w04	[1; 20; 1]	µm	[100 - 300]	>100	mV
M20 (_w04, _l04, 1)	_w10	[1; 20; 1]	µm			
M21 (_w11, _l11, 1)	_w11	[1; 20; 1]	µm			
M3a = M3b (_w02, _l03, m03)	_l00	[0.35; 10; 0.05]	µm	[100 - 300]	>100	mV
M5a = M5b (_w04, _l05, m05)	_l01	[0.35; 10; 0.05]	µm	[100 - 300]	>100	mV
M6a = M6b (_w02, _l06, m06)	_l02	[0.35; 10; 0.05]	µm	[50 - 300]	>100	mV
	_l03	[0.35; 10; 0.05]	µm			
Bias Circuit	_l04	[0.35; 10; 0.05]	µm			
M16=M17 (_w02, _l02, 1)	_l05	[0.35; 10; 0.05]	unit	[100 - 300]	>100	mV
M18 (_w10, _l10, 1)	_l06	[0.35; 10; 0.05]	µm			
M19 (_w02, _l03, 1)	_l10	[0.35; 10; 0.05]	µm			
M4a=M4b (_w04, _l04, m04)	_l11	[0.35; 10; 0.05]	µm	[100 - 300]	>100	mV
M7a=M7b (_w04, _l05, m05/2)	_m01	[1; 80;1]	unit	[100 - 300]	>100	mV
M1a=M1b (_w01, _l01, m01)	_m02	[1; 80;1]	unit	[50 - 300]	>100	mV
	_m03	[1; 80;1]	unit			
	_m04	[1; 80;1]	unit			
	_m05	[1; 80;1]	unit			
	_m06	[1; 80;1]	unit			

(a) Technology Constraints - overdrive voltages (b) Drain-sources voltages

Table 6.24 Optimization algorithm parameters

Parameter	**value**	**Parameter**	**value**	**Parameter**	**value**
Kernel	GA-MOD	**Selection**	Tournament	**Popsize**	64
Strategy	Corner Optimization	**Crossover**	Two point	**Init Pop**	2*Popsize
Sampling	LHS	**Mutation**	Dynamic	**Generations**	150
Sort	Priority to performance fitness then perf. constraints.	**Adaptive**	No	**Stop condition**	End of generations
		Elite	25% of populat.	**Search** Space	2,370e+37

6.5.1.2 Analysis

The attached test benches used for DC and AC simulations are illustrated in Fig. 6.12. The unusual values for the resistance (1T Ohm) and for the capacitance (1F) ensure the same voltage in DC Analysis for nodes *Vin-*, *Vin+*, *Voutp* and *Voutn*, it is also possible to analyze the amplifier open loop gain. A dependent source voltage is used to transform a differential output (*voutp*, *voutn*) into a single ended one (*voutd*).

The simulation results for the main amplifier and *cmfb* circuit sizing achieved with the optimization module, and using the HSPICE simulator as the evaluation engine are presented in Table 6.25 and satisfy all the design requirements. The final transistor dimensions are displayed in Table 6.26. The proper biasing of all CMOS transistors are guaranteed once the final solution satisfies all the design specs and functional constraints for each of the corner points. The computational times were included, in Table 6.27, to illustrate the effectiveness of the proposed system.

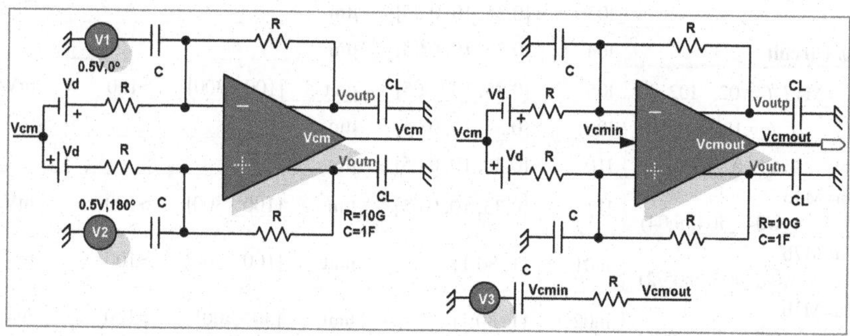

Fig. 6.12 Testbench for (a) AC and (b) AC Common mode feedback specifications

Table 6.25 Performance parameter specifications

	Specifications	Target	Sizing Result	Units
Electrical	GBW	> 100	158.0	MHz
	Phase margin	> 60	65.0	o
	DC gain	> 55	66.6	dB
	CMF GBW	> 50	64.1	MHz
	CMF Phase margin	> 60	75.6	o
Optimization	Power Consumption	Minimum	4.2	mW
	Current Consumption	Minimum	1.2	mA

Table 6.26 Final transistor dimensions

Main Amplifier	W/L (μm/μm)	Bias	W/L (μm/μm)	Common Mode	W/L (μm/μm)
M0	54/0.40	M0a, b, c	1/0.40	M6a, b	170/0.95
M1a, M1b	41/0.95	M16,17	10/0.40	M7a, b	45/1.75
M2a, M2b	90/0.40	M18	3/0.95	M2c	90/0.40
M3a, M3b	170/0.90	M19	10/0.90		
M4a, M4b	441/7.80	M20	9/7.80		
M5a, M5b	54/0.95	M21	6/4.40		

Table 6.27 Runtime info*

Design Problem			
Opt. Variables / Constraints (Specs + Design Const.)	21	7 + 34 = 41	
(1 -Step) Corners Optimization	**Time**	**#Generation**	**#Evaluations**
Overall Optimization time	17m05s	150	2496
First Feasible Solution	10m56s	79	1329
Best feasible solution	12m45s	92	1616

* In a single processor Intel(R) Core(TM)2 Quad CPU Q6600 @ 2.40GHz PC running Linux.

The next pages show all graphical and numerical results for the AC corner analysis. Fig. 6.13 shows all the gain magnitudes, it is interesting to observe the range of DC gain and GBW; all corner numerical results are reported in Table 6.28 while Table 6.29 shows minimum and maximum values.

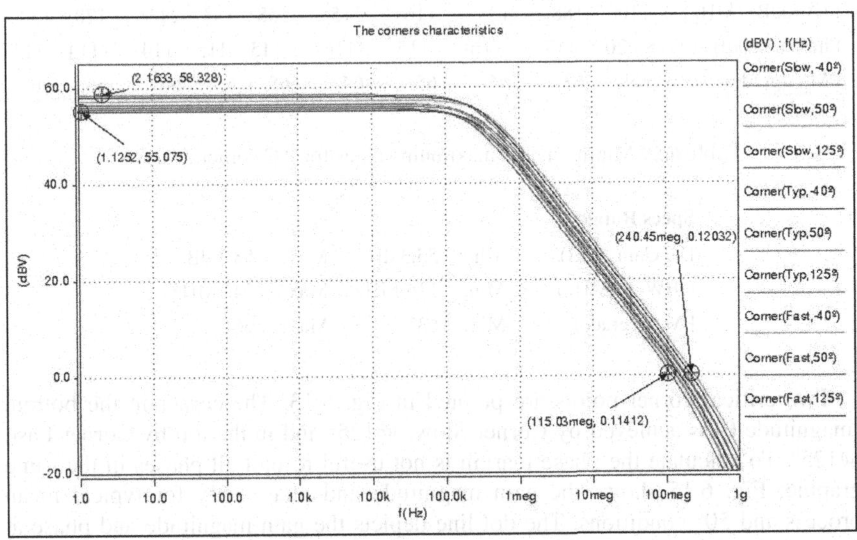

Fig. 6.13 Gain magnitudes for corners analysis

Fig. 6.14 displays the output from corner simulation.

CORNER	MODEL	TEMP	WEIGHT	SATISFY	FITNESS(i)	VIOL(i)	SUM_FIT[10150]
[1]	CSLOW	-40º	1	7 / 34=41	6.779e-03	0.000e+00	6.779e-03
[2]	CSLOW	+50º	1	7 / 34=41	1.333e-02	0.000e+00	1.333e-02
[3]	CSLOW	+120º	1	7 / 34=41	1.980e-02	0.000e+00	1.980e-02
[4]	CTYP	-40º	1	7 / 34=41	3.438e-02	0.000e+00	3.438e-02
[5]	CTYP	+50º	1	7 / 34=41	4.830e-02	0.000e+00	4.830e-02
[6]	CTYP	+120º	1	7 / 34=41	6.199e-02	0.000e+00	6.199e-02
[7]	CFAST	-40º	1	7 / 34=41	7.012e-02	0.000e+00	7.012e-02
[8]	CFAST	+50º	1	7 / 34=41	7.774e-02	0.000e+00	7.774e-02
[9]	CFAST	+120º	1	7/ 34=41	8.518e-02	0.000e+00	8.518e-02

****** EUREKA ******

Byebye. AIDA - IC_DESIGN Terminate ...

Job done on a Intel(R) Core(TM)2 Quad CPU Q6600 @ 2.40GHz

Fig. 6.14 Output from simulation where all corners are satisfied

Table 6.28 Results for corners analysis

Corner		**1**	**2**	**3**	**4**	**5**	**6**	**7**	**8**	**9**
Process		slow			typical			fast		
Temperature		-40°	50°	125°	-40°	50°	125°	-40°	50°	125°
Specs	**Values**									
DC Gain (dB)	> 55	58.3	56.5	55.5	57.9	56.3	55.4	57.2	55.9	55.3
f (A=0dB) (MHz)	> 100	188	138	116	215	158	133	243	178	150
Phase (A=0dB) (°)	>-120	-117	-116	-115	-116	-115	-115	-114	-114	-114
PM (grade)	> 60	63	64	65	64	65	65	66	66	66

Table 6.29 Minimum and maximum values for AC corner analysis

Specs Range				
DC Gain (dB)	Min:	55.3 dB	Max:	58.3 dB
GBW (MHz)	Min:	116 MHz	Max:	243 MHz
PM (grade)	Min:	63°	Max:	66°

Two critical corner points are pointed in Fig. 6.13. The corner in the bottom (magnitude 0) is achieved by Corner Slow, @125° and in the top by Corner Fast, @125°. To calculate the phase margin is not useful to plot all phases in the same graphic; Fig. 6.15 shows the gain magnitude and phase only for typical mean process and 50° conditions. The dot line depicts the gain magnitude and phase at the common mode output.

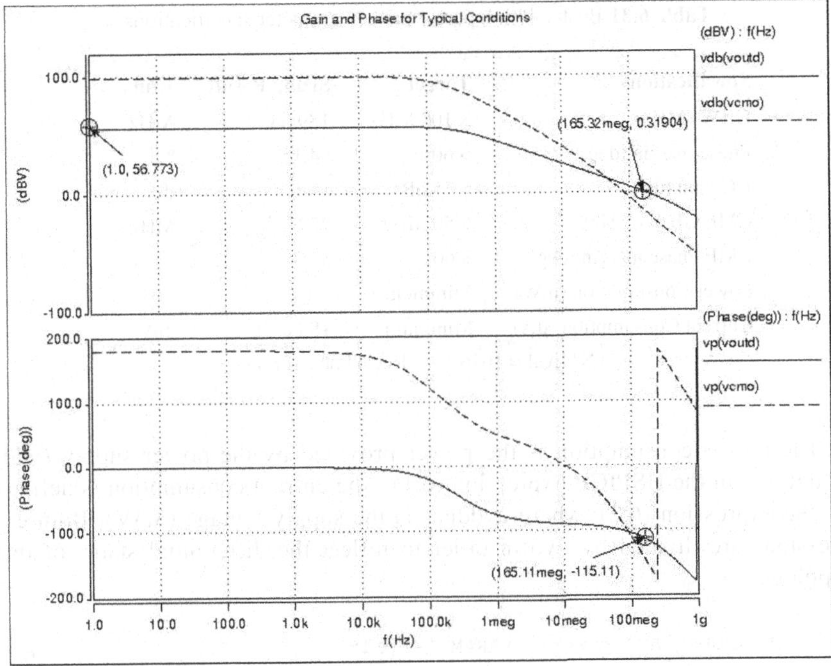

Fig. 6.15 Gain magnitude and phase for typical conditions

6.5.1.3 Design Analysis

The GENOM optimization algorithm solves the circuit sizing problem with efficiency considering the type of optimization evolved in this experiment, the corner optimization and taking also in consideration the number of optimization variables and constraints. The first and the final solutions produced are presented in Table 6.30 and Table 6.31.

Table 6.30 First feasible solution performance parameter specifications

Specifications	Target	Sizing Result	Units
GBW (MHz)	> 100 MHz	110.6	MHz
Phase margin (deg)	> 60º	74.0	º
DC gain (dB)	> 55 dB	+61.1	dB
CMF GBW (MHz)	> 50 MHz	60.7	MHz
CMF Phase margin (deg)	> 60º	+82.4	º
Power Consumption (mW)	Minimum	2.8	mW
Current Consumption (μA)	Minimum	8.7e-01	mA

Nº Eval = 1329 RealTime: 10min

Table 6.31 Best solution performance parameter specifications

Specifications	Target	Sizing Result	Units
GBW (MHz)	> 100 MHz	189.38	MHz
Phase margin (deg)	> 60º	64.1	º
DC gain (dB)	> 55 dB	58.1	dB
CMF GBW (MHz)	> 50 MHz	75.6	MHz
CMF Phase margin (deg)	> 60º	84.8	º
Power Consumption (mW)	Minimum	2.1	mW
Current Consumption (μA)	Minimum	+8.8e-01	mA

Nº Eval = 1626 RealTime: 12min

The power consumption is the power provided by the power supply (vdd) as defined in the HSPICE expression (6.1). The current consumption is defined by the expression (6.2), where avddpar is the supply voltage (3.3V). Both expressions are divided by two in order to reflect the differential status of this topology.

```
.MEASURE  AC 'power'  PARAM('-P(vdd)/2')
```
(6.1)

```
.MEASURE AC 'iavdd'  PARAM('-P(vdd)/avddpar/2')
```
(6.2)

6.5.2 A Common OTA Fully Differential Telescopic OpAmp

6.5.2.1 Description

A common OTA (Operational Transconductance Amplifier) is the telescopic amplifier. The major drawback of this amplifier's topology is the reduced output swing when compared with other solutions, such as the folded cascade or two stage amplifiers, which becomes relevant in low voltage applications. On the other hand, its good speed performance associated with its low power consumption turns this topology into a competitive implementation. The schematic represented in Fig. 6.16 is an in-house fully differential version of this topology.

The topology consists in 24 transistors grouped in 2 main functional blocks: the main amplifier with differential input and the bias circuit. A quick inspection to circuit schematic highlights the potential matching of some groups of transistors like *M0* and *M19*, *M40* and *M43*, *M17* and *M18*, *M34* and *M36*. Some dependent relations like for instance, the multiplicity factor of transistors *M18*, *M17*, *M24* and *M5* and others listed in Table 6.33 should also be checked.

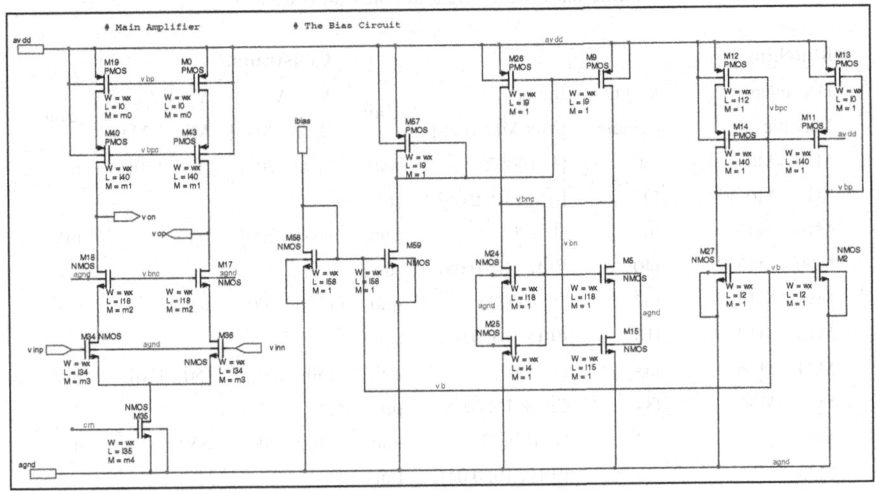

Fig. 6.16 Telecopic OpAmp - Main amplifier and Bias circuitry

6.5.2.2 Problem Specifications and Design Configurations

The main objective was to synthesize the presented telescopic amplifier, using the *UMC 0.18μm* logic 1.8V Generic II process, according to the performance specifications listed in Table 6.32, and designed to follow the fundamental designer rules and optimization design constraints of Table 6.33. The total number of constraints, performance constraints and the constraints derived from designer's rules are composed by 23 optimizations constraints that must be satisfied for the optimization process described in Table 6.35. The specifications must be satisfied for the corners points of Table 6.36.

Table 6.32 Performance parameter specifications

	Specifications	Target	Units	Description
Electrical	DC gain	> 75	dB	DC gain
	GBW	> 100	MHz	Unit-gain frequency
	Phase Margin	[60-90]	o	Phase margin
Environmental	Capacitive Load	1.1	pF	Capacitive Load
	Wi – fixed widths	2	μm	Fixed all widths
Optimization	Power Consumption	Minimum	mW	Objective
	Current Consumption	Minimum	μA	Objective

Table 6.33 Matching and constraints details

Matching				**Constraints**		
Dependent Variable	Design Variable	Range [Min Max;Step]	Unit	V_{GS} - V_T [(a)] [Min - Max]	V_{DS} - V_{DSAT} [(b)] Min / Max	Unit
M19 – M0	_m0	[1; 100; 2]	unit	[100 - 200]	[50 - 150]	mV
M19 – M0	_l0	[0.18; 10; 0.05]	μm			
M40 – M43	_m1	[1; 100; 2]	unit	[100 - 200]	>50	mV
M40 – M43	_l40	[0.18; 10; 0.05]	μm			
M18 – M17	_m2	[1; 100; 2]	unit	[100 - 200]	>50	mV
M18 – M17	_l18	[0.18; 10; 0.05]	μm			
M34 – M36	_m3	[1; 100; 2]	unit	[50 - 200]	[50 - 150]	mV
M34 – M36	_l34	[0.18; 10; 0.05]	μm			
M35	_m4	[1; 100; 2]	unit	[100 - 200]	>50	mV
M35	_l35	[0.18; 10; 0.05]	μm			
Bias Circuit						
M24 – M5	_l18	[0.18; 10; 0.05]	μm			
M59 – M58	_l58	[0.18; 10; 0.05]	μm			
M9 – M57 – M26	_l19	[0.18; 10; 0.05]	μm			
M11 – M14	_l40	[0.18; 10; 0.05]	μm			
M13	_l0	[0.18; 10; 0.05]	μm			
M12	_l12	[0.18; 10; 0.05]	μm			
M27 – M2	_l2	[0.18; 10; 0.05]	μm			
M15	_l15	[0.18; 10; 0.05]	μm			
M25	_l25	[0.18; 10; 0.05]	μm			

(a) Technology Constraints - overdrive voltages (b) Drain-sources voltages

Table 6.34 explains the rationale behind the achieved constraints values used in this experiment. In a fully differential amplifier, as the one shown in Fig. 6.16, the amplifier can be designed in two symmetrical parts. When one transistor changes value, its mirror also changes. This principle is used for the input differential pair, the cascode and load transistors. As for the overdrive voltage and margin, the constraints are as follows:

Table 6.34 Matching and constraints details

	Overdrive voltage Vgs-Vt = Vov	**Margin** Vds-VDsat
differential pair	50mV> Vov >200mV	> 50mV
current sources	50mV>Vov>200mV	> 50mV
cascodes	50mV>Vov>200mV	> 50mV
current sources with cascodes	50mV>Vov>200mV	50mV> Margin > 200mV

Table 6.35 Optimization algorithm parameters

Parameter	**Value**	**Parameter**	**value**	**Parameter**	**value**
Kernel	GA-MOD	**Selection**	Tournament by "feasibility"	**Popsize**	64
Strategy	Typical + Corner Optimization	**Crossover**	Two point	**Init Pop**	2*Popsize
Sampling	LHS	**Mutation**	Dynamic	**Generations**	150
Sort	Priority to performance fitness then performance constraints.	**Adaptive**	No	Stop	End of generations
		Elite	25% of population	**Search Space domain**	2.344e+35

Table 6.36 Corner analysis data

Conditions	**Variation points**		
MOS worst case parameters	SF-Slow	TT-Typ	FS-Fast
Temperature Range (° C)	-40° C	+50° C	+120°C

Where, SF, TT and FS means the Slow/Fast, Typical/Typical and Fast/Slow process, respectively. Instead of using the typical fast and slow device models sets, where all devices are supposed to be fast or slow, a mixture of slow nMOS devices and fast pMOS is here considered, for example purposes, namely the SF, TT and FS meaning the Slow/Fast, Typical/Typical and Fast/Slow process, respectively.

6.5.2.3 Analysis

The attached test bench circuit used for DC and AC simulations is illustrated in Fig. 6.17. A dependent source voltage is used to transform a differential output (*out1*, *out2*) into a single ended one (*outd*).

This experiment was executed on a single Intel(R) Core(TM)2 Quad CPU Q6600 @ 2.40GHz dual core machine and use HSPICE to simulate the circuit and extract performance parameters. The simulation results of the main amplifier and bias circuit sizing are shown in Table 6.37. The final transistor dimensions for all the devices and biasing conditions resulting from the sizing process are displayed in Table 6.38.

Fig. 6.18 shows the gain magnitude and phase for typical process and 50°C conditions.

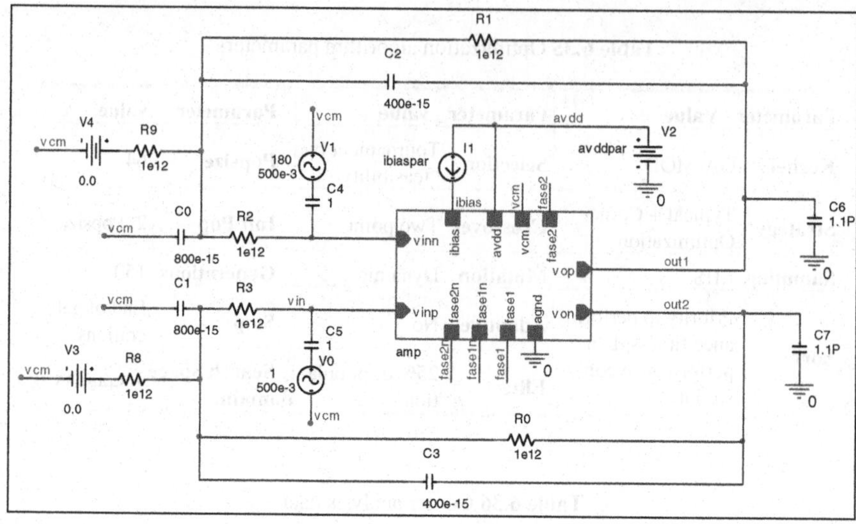

Fig. 6.17 Telescopic OpAmp - Testbench for DC and AC specifications

Table 6.37 Performance parameter specifications

	Specifications	Target	Sizing Result	Units
Electrical	DC gain	> 75	77.6	dB
	GBW	> 100	123.0	MHz
	Phase margin	> 60	65.0	o
Optimization	Power Consumption	Minimum	5.6e-01	mW
	Current Consumption	Minimum	3.1e-01	mA

Table 6.38 Final transistor dimensions

Main Amplifier	W/L (μm/μm)	Bias	W/L (μm/μm)
M19 – M0	202 / 1.080e-06	M24 – M5	2 / 1.33e-06
M40 – M43	152 / 1.58e-06	M59 – M58	2 / 1.38e-06
M18 – M17	126 / 1.33e-06	M9 – M57 – M26	2 / 4.03e-06
M34 – M36	60 / 0.73e-06	M11 – M14	2 / 1.58e-06
M35	10 / 0.18e-06	M13	2 / 1.08e-06
		M12	2 / 8.53e-06
		M27 – M2	2 / 8.63-06
		M15	2 / 1.73e-06
		M25	2 / 9.68e-06

Note: M0c belongs to Bias and have the same value that M0a and M0b.

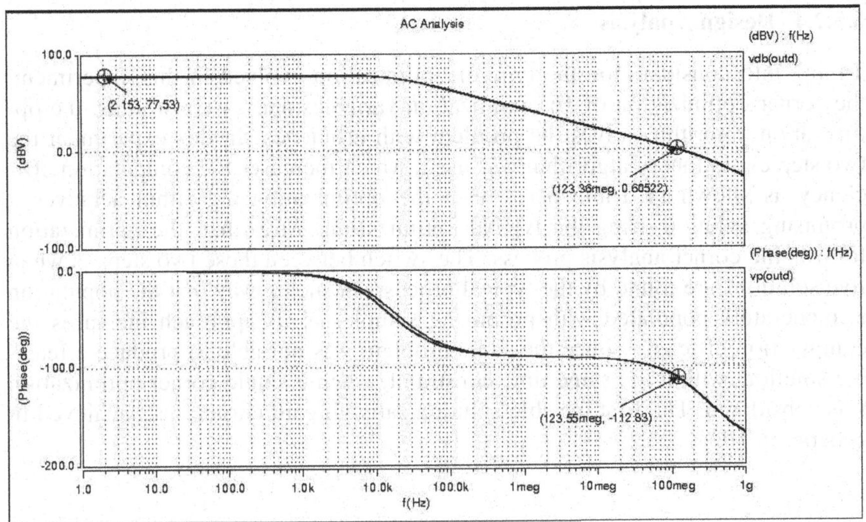

Fig. 6.18 Gain magnitude and phase for typical conditions

As it can be noticed, this simulation design meets the required specs related to DC gain, gain bandwidth and phase margin satisfying all corners points as reported in Table 6.39, while Table 6.40 shows the minimum and maximum values. Obviously the amplifier was designed in order to obtain a worst case *DC gain* bigger than 75dB and a *GBW* bigger than 110MHz.

Table 6.39 Numerical results for corner analysis

Corner		1	2	3	4	5	6	7	8	9
Process		Slow-Fast			Typical			Fast-Slow		
Temperature		-40°	50°	125°	-40°	50°	125°	-40°	50°	125°
Specs	**Values**									
DC Gain (dB)	> 55	78.3	76.5	75.5	78.9	77.6	76.4	77.2	75.9	75.1
f (A=0dB) (MHz)	> 100	178	128	110	205	123	133	223	146	140
Phase (A=0dB) (°)	>-120	-117	-116	-115	-116	-115	-115	-114	-114	-114
PM (grade)	> 60	63	64	65	64	65	65	66	66	66

Table 6.40 Minimum and maximum values for AC corner analysis

Specs Range				
DC Gain (dB)	Min:	75.1 dB	Max:	78.9 dB
GBW (MHz)	Min:	110 MHz	Max:	223 MHz
PM (grade)	Min:	63°	Max:	66°

6.5.2.4 Design Analysis

Taking into consideration the type of optimization evolved in this experiment, the corner optimization of 16 optimization variables and 24 constraints, the optimization algorithm solves the problem with efficiency. In this experiment the two step evolutionary algorithm was used, which increases the computation efficiency as shown in Table 6.41, once the optimization algorithm achieves a promising solution using the typical optimization. After that, the optimization follows the corner analysis process. The switch between these two steps is when five solutions are found by the typical process, in such a way that the population is moderately populated with promising samples. This approach increases the computation efficiency once the same problem was not able to produce a feasible solution, within the same time constraint, when a single corner optimization was considered. The first feasible solution satisfying all corners was achieved in generation 102.

Table 6.41 Runtime info

Design Problem			
Opt. Variables / Constraints (Specs + Design Const.)	16	5 + 18 = 24	
(1-Step) Typical Optimization	**Time**	**#Generation**	**#Evaluations**
Overall Optimization time	2m12s	53	944
First Feasible Solution	1m00s	22	427
Best feasible solution	1m48s	37	661
(2 -Step) Corners Optimization	**Time**	**#Generation**	**#Evaluations**
Overall Optimization time	14m09s	150	3184
First Feasible Solution	10m06s	102	1664
Best feasible solution	12m07s	120	1937

* In a single processor Intel(R) Core(TM)2 Quad CPU Q6600 @ 2.40GHz PC running Linux.

The resolution for this problem was achieved using an iterative process very similar to the traditional analog design. In a first attempt to solve the problem, it was observed that one of the corners in particular was very difficult to satisfy. This corner was identified by the inspection of the run-time information returned from simulation and provided by the tool. This critical corner point (corner nº3) is pointed in Fig. 6.19. The simulation was interrupted and the static weight for that corner was changed as shown in Fig. 6.20, and the simulation was rerun again. Finally it was possible to obtain several solutions within the original time constraint.

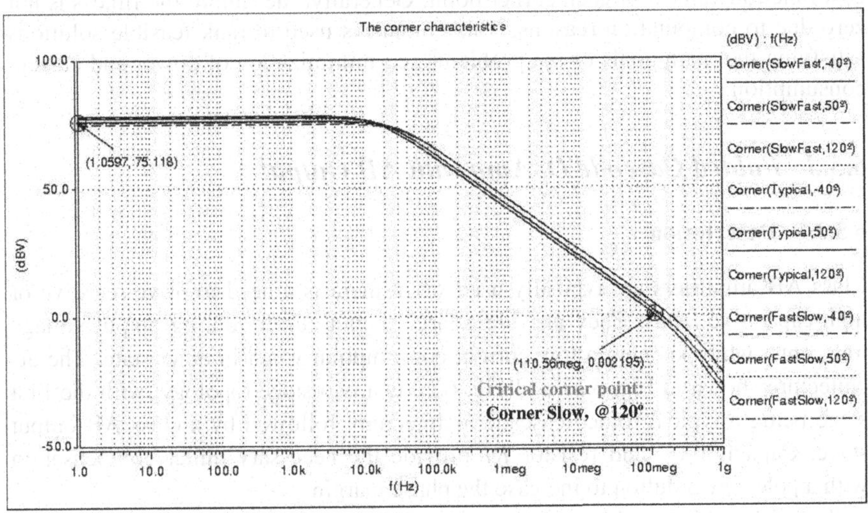

Fig. 6.19 Gain magnitudes for corner analysis

CORNER	MODEL	TEMP	WEIGHT	SATISFY	FITNESS(i)	VIOL(i)	SUM_FIT[23104]
[1]	SF-SLOW	-40°	1	5 / 18=23	1.196e-03	0.000e+00	1.196e-03
[2]	SF-SLOW	+50°	1	5 / 18=23	2.343e-03	0.000e+00	3.540e-03
[3]	*SF-SLOW*	+120°	2	5 / 18=23	3.592e-02	0.000e+00	3.946e-02
[4]	TT-TYP	-40°	1	5 / 18=23	5.782e-03	0.000e+00	4.524e-02
[5]	TT-TYP	+50°	1	5 / 18=23	3.923e-03	0.000e+00	4.916e-02
[6]	TT-TYP	+120°	1	5 / 18=23	2.040e-02	0.000e+00	6.957e-02
[7]	FS-FAST	-40°	1	5 / 18=23	2.224e-02	0.000e+00	1.418e-01
[8]	FS-FAST	+50°	1	5 / 18=23	4.035e-02	0.000e+00	1.821e-01
[9]	FS-FAST	+120°	1	5 / 18=23	5.146e-02	0.000e+00	2.336e-01

****** EUREKA ******

Byebye. AIDA - IC_DESIGN Terminate ...

Job done on a Intel(R) Core(TM)2 Quad CPU Q6600 @ 2.40GHz

Fig. 6.20 Output from simulation where all corners are satisfied

All parameters from column "*VIOL*" have null values indicating the constraints related to designer's rules (18) were totally satisfied. Additionally, the column "*SATISFY*" confirms that all constraints including the performances (5+18) were satisfied in all corner points. The column "*FITNESS*" represents the cost function values for each corner point. Generally, the sum of the fitness is not zero due to computation reasons. This amount is used to rank feasible solutions satisfying the main goals of the problem, i.e., minimization of power and current consumption.

6.5.3 Folded Cascode OpAmp with AB Output

6.5.3.1 Description

Class AB amplifiers are typically used when there is a need to drive resistive or high capacitive loads. They provide a large output current during output voltage transients, while keeping a low current consumption when in quiet state. The architecture shown in Fig. 6.21 and Fig. 6.22 is a two stage topology, with the first stage being a typical folded cascade architecture, followed by a class AB output stage. Capacitor *C1* and resistor *R1* provide the necessary miller compensation with a pole zero solution to increase the phase margin.

Transistors *M6* together with transistors *M22* to *M27* provide the control of the class AB operation by controlling the maximum output current of *M7* and *M8*. This control is performed by keeping control of the *Vgs* voltage of *M6a* and *M8* so that *Vgs6a* + *Vgs8* = *Vgs24* + *Vgs27*. Therefore the maximum output current supplied by *M8* is controlled by the current in *M24* and *M27*. For positive currents the same principle is applied to *M6b*, *M7*, *M23* and *M25*.

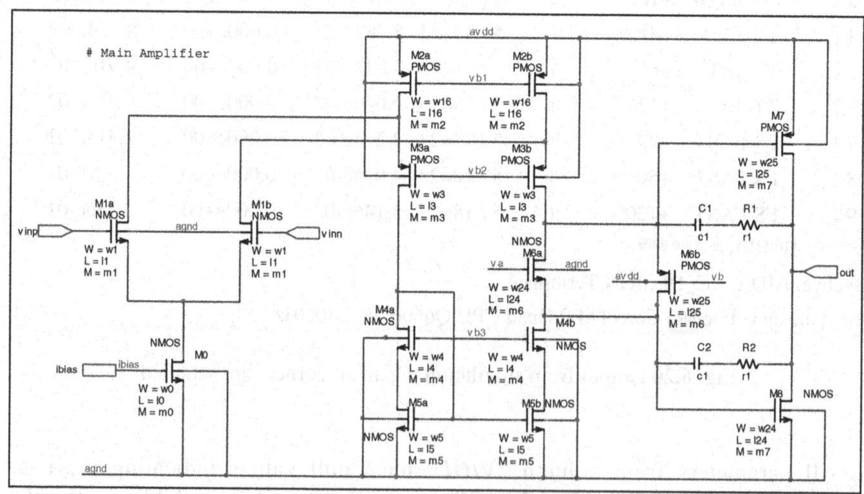

Fig. 6.21 Main class AB Amplifier

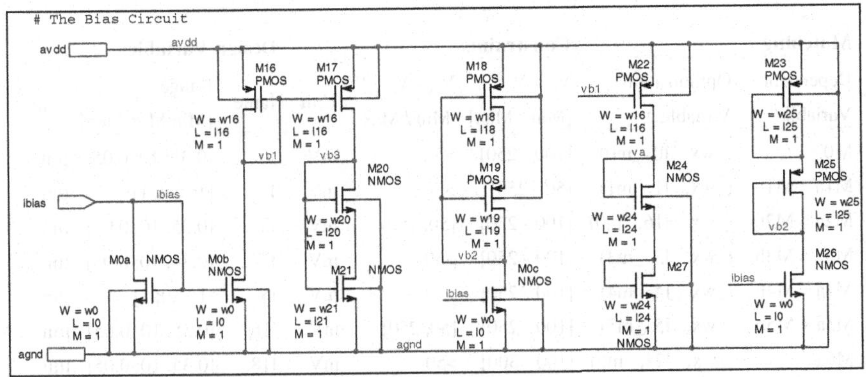

Fig. 6.22 Bias circuit

6.5.3.2 Problem Specifications and Design Configurations

The main objective was to synthesize the presented folded cascode amplifier using the AMS (Austria Mikro Systeme Intl. AG) 0.35 μm, 3.3 V CMOS technology according to the performance specifications listed in Table 6.42, and designed to follow the fundamental designer rules and optimization design constraints of Table 6.43. The specifications must be satisfied for the corner points of Table 6.44. The total of constraints (performance constraints and the constraints derived from designer's rules) results in 33 optimizations constraints that must be satisfied in the optimization process described in Table 6.45.

Table 6.42 Performance parameter specifications

	Specifications	Target	Units	Description
Electrical	gain_dc	> 70	dB	Unit-gain frequency
	gbw	> 75	MHz	Phase margin
	phase	[60-90]	º	DC gain
Environmental	CL	1	pF	Capacitive Load
	Ibiaspar	10	μA	Ibias
	Wi fixed	5	μm	Fixed all widths
Optimization	Power Consumption	Minimum	mW	Objective
	Current Consumption	Minimum	μA	Objective

Table 6.43 Matching and constraints details

Matching		**Constraints**			**Design Variable**		
Dependent Variable	Optimization Variable	$V_{GS} - V_T$ [(a)] [Min - Max]	$V_{DS} - V_{DSAT}$ [(b)] Min / Max	Unit	Name	Range [Min Max;Step]	Unit
M0	(_wx, _l0, _m0)	[100 - 250]	>50	mV	_l0	[0.35; 10; 0.05]	µm
M1a = M1b	(_wx, _l1, _m1)	[50 - 250]	>50	mV	_l1	[1; 20; 1]	µm
M2a = M2b	(_wx, _l16, _m0)	[100 - 250]	[50, 250]	mV	_l3	[0.35; 10; 0.05]	µm
M3a = M3b	(_wx, _l3, _m3)	[100 - 250]	>50	mV	_l4	[0.35; 10; 0.05]	µm
M4a = M4b	(_wx, _l4, _m4)	[100 - 250]	>50	mV	_l5	[1; 20; 1]	µm
M5a = M5b	(_wx, _l5, _m5)	[100 - 250]	[50, 250]	mV	_l16	[0.35; 10; 0.05]	µm
M6a	(_wx, _l24, _m6)	[100 - 300]	>50	mV	_l18	[0.35; 10; 0.05]	µm
M6b	(_wx, _l25, _m6)	[100 - 300]	>50	mV	_l21	[0.35; 10; 0.05]	µm
M7	(_wx, _l25, _m7)	[100 - 250]	>50	mV	_l24	[0.35; 10; 0.05]	µm
M8	(_wx, _l24, _m7)	[100 - 250]	>50	mV	_l25	[0.35; 10; 0.05]	µm
R1=R2	_r1				_m0	[1; 100;1]	unit
C1=C2	_c1				_m1	[1; 100;1]	unit
Bias Circuit:					_m3	[1; 100;1]	unit
Dep.Variable	Opt.Variable	Dep.Variable	Opt.Variable		_m4	[1; 100;1]	unit
M20	(_wx, _l4, 1)	M16 =M17	(_wx, _l16, 1)		_m5	[1; 100;1]	unit
M21	(_wx, _l21, 1)	M18	(_wx, _l18, 1)		_m6	[1; 100;1]	unit
M23 =M25	(_wx, _l25, 1)	M19	(_wx, _l3, 1)		_m7	[1; 100;1]	unit
M24 =M27	(_wx, _l24, 1)	M0a=M0b== M0c	(_wx, _l0, 1)		_r1	[100;1000;50]	Ω
M26	(_wx, _l0, 1)				_c1	[1;5;0.05]	tF

(a) Technology Constraints - overdrive voltages (b) Drain-sources voltages

Table 6.44 Corners analysis data

Conditions	**Variation points**		
MOS worst case parameters	Ws-Slow	Tm-Typ	Wp-Fast
Temperature Range (° C)	-40° C	+50° C	+120°C

Table 6.45 Optimization algorithm parameters

Parameter	**value**	**Parameter**	**value**	**Parameter**	**value**
Kernel	GA-MOD	**Selection**	Tourn. by feas.	**Popsize**	64
Strategy	Corner Optim.	**Crossover**	Two point	**Init Pop**	2*Popsize
Sampling	LHS	**Mutation**	Dynamic	**Generations**	250
Sort	Priority to performance fitness then performance constraints.	**Adaptive**	Yes	**Stop**	End of generations
		Elite	25% of pop.	**Search Space domain**	6.417e+39

Note: Evaluation Engine by HSPICE simulator.

6.5.3.3 Design Analysis

The attached testbench circuit used for DC and AC simulations is illustrated in Fig. 6.23.

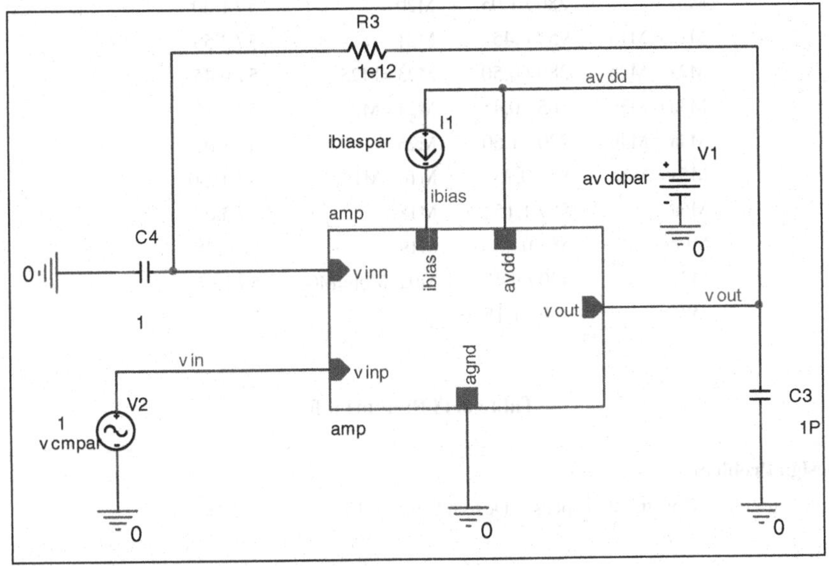

Fig. 6.23 OpAmp testbench for DC and AC specifications

The simulation results of the main amplifier and bias circuit sizing are shown in Table 6.46. The final transistor dimensions are displayed in Table 6.47, while, Table 6.48 summarizes the runtime information for this one step corner optimization.

Fig. 6.24 gives an outline of the text simulation data produced by the optimization tool of one feasible solution.

All parameters from column "VIOL" have null values indicating the constraints related to designer's rules (28) were totally satisfied. Additionally, the column "SATISFY" confirms that all constraints (5+28) were satisfied in all corner points. The column "FITNESS" represents the fitness values for each corner point. Generally, the sum of the fitness is not zero due to computation reasons. This amount is used to rank feasible solutions satisfying the goals of the problem.

Table 6.46 - Performance parameter specifications

	Specifications	Target	Sizing Result	Units
Electrical	DC gain	> 70	94.7	dB
	GBW	> 75	115.1	MHz
	Phase margin	[60-90]	69.0	o
Optimization	Power Consumption	Minimum	6.1	mW
	Current Consumption	Minimum	1.8	mA

Table 6.47 Final transistor dimensions

Main Amplifier	W/L (μm/μm)	**Bias**	W/L (μm/μm)
M0	280 / 1.05	M20	5 / 1.60
M1a = M1b	95 / 0.45	M21	5 / 3.55
M2a = M2b	280 / 0.50	M23 =M25	5 / 0.45
M3a = M3b	165 / 0.45	M24 =M27	5 / 1.15
M4a = M4b	220 / 1.60	M26	5 / 1.05
M5a = M5b	85 / 0.75	M16 =M17	5 / 0.50
M6a	55 / 1.15	M18	5 / 3.05
M6b	55 / 0.45	M19	5 / 0.45
M7	430 / 0.45	M0a=M0b=M0c	5 / 1.05
M8	430 / 1.15		

Table 6.48 Runtime info

Design Problem		
Opt. Variables / Constraints (Specs + Design Const.)	19	5 + 28 = 33

(1 -Step) Corners Optimization	**Time**	**#Generation**	**#Evaluations**
Overall Optimization time	23m04s	250	4061
First Feasible Solution	20m03s	227	3702

* In a single processor Intel(R) Core(TM)2 Quad CPU Q6600 @ 2.40GHz PC running Linux.

CORNER	MODEL	TEMP	WEIGHT	SATISFY	FITNESS(i)	VIOL(i)	SUM_FIT[1130]
[1]	WS-SLOW	-40º	1	5 / 28=33	3.845e-03	0.000e+00	3.845e-03
[2]	WS-SLOW	+50º	1	5 / 28=33	6.113e-03	0.000e+00	6.113e-03
[3]	WS-SLOW	+120º	1	5 / 28=33	8.628e-03	0.000e+00	8.628e-03
[4]	Tm-TYP	-40º	1	5 / 28=33	1.570e-02	0.000e+00	1.570e-02
[5]	Tm-TYP	+50º	1	5 / 28=33	3.238e-02	0.000e+00	3.238e-02
[6]	Tm-TYP	+120º	1	5 / 28=33	4.600e-02	0.000e+00	4.600e-02
[7]	Wp-FAST	-40º	1	5 / 28=33	5.720e-02	0.000e+00	5.720e-02
[8]	Wp-FAST	+50º	1	5 / 28=33	7.653e-02	0.000e+00	7.653e-02
[9]	Wp-FAST	+120º	1	5 / 28=33	8.840e-02	0.000e+00	8.840e-02

****** EUREKA ******

Byebye. AIDA - IC_DESIGN Terminate ...

Job done on a Intel(R) Core(TM)2 Quad CPU Q6600 @ 2.40GHz

Fig. 6.24 Output from simulation where all corners are satisfied

Fig. 6.26 shows the gain magnitude and phase only for typical mean process and 50° C conditions. Fig. 6.26 shows the graphical results for the AC corner analysis. As it can be noticed, this simulation design meets the required specs related to DC gain, gain bandwidth and phase margin satisfying all corner points as reported in Table 6.49. Table 6.50 shows the maximum and minimum of the corner points.

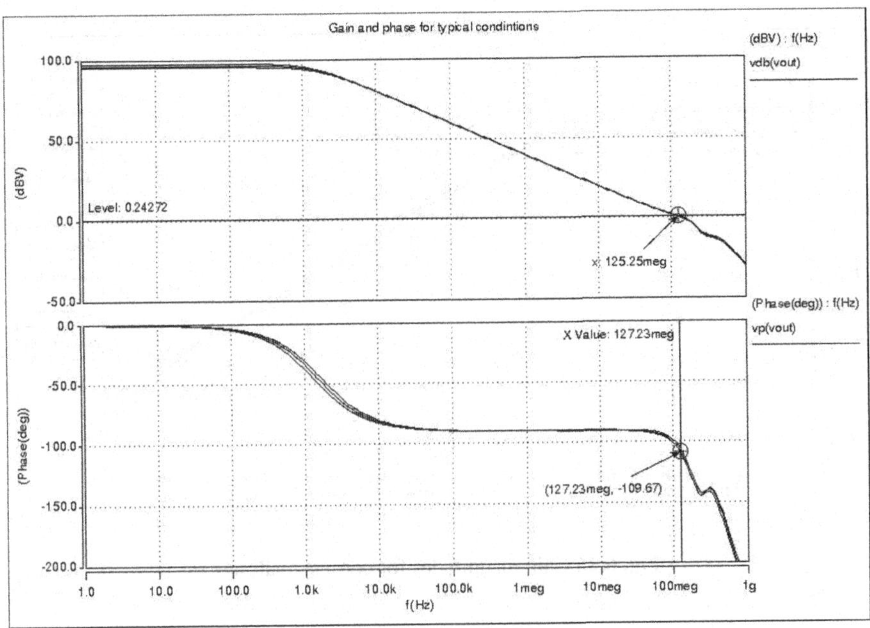

Fig. 6.25 Gain magnitude and phase for typical conditions

Table 6.49 Numerical results for corners analysis

Corner		1	2	3	4	5	6	7	8	9
Process			Slow			Typical			Fast	
Temperature		-40°	50°	125°	-40°	50°	125°	-40°	50°	125°
Specs	**Values**									
DC Gain (dB)	> 70	102.1	98.9	97.6	96	94.6	92.7	89.1	87.3	84.4
f (A=0dB) (MHz)	> 75	131	101.7	100.5	173.4	115.1	104.5	246	158	109
Phase (A=0dB) (°)	------	-115	-118	-119	-112	-111	-112	-110	-102	-101
PM (grade)	[60-90]	64	61	60	67	68	67	69	77	76

Table 6.50 Minimum and maximum values for AC corners analysis

Specs Range				
DC Gain (dB)	Min:	84,4 dB	Max:	102.1 dB
GBW (MHz)	Min:	100.5 MHz	Max:	246 MHz
PM (grade)	Min:	60º	Max:	77º

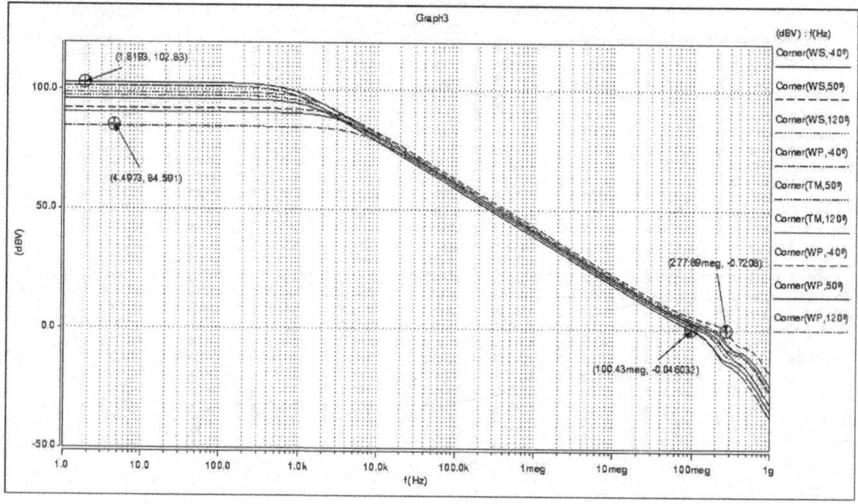

Fig. 6.26 Gain magnitudes for corners analysis

6.6 Comparison with Other Tools/Approaches

The lack of a known open reference tool for IC design automation makes it difficult to the evaluation task of comparing objectively different implementations, although, the analog design automation community is developing efforts to circumvent this situation. Comparing the performance and effectiveness of the final GENOM optimizer with published reference tools is not always possible because the information contained in most of the publications omit some detail of the implementation, maybe imposed by logistics limitations or by author intentionality focusing only the most important piece of interest. Some common ignored items are related with incomplete definition of testbench circuitry, range of optimization variables, used device models and insufficient output data exposed. An exception is made for the first benchmark circuit presented above that was gently provided by Prof. Francisco Fernandez, IMSE-CNM-CSIC/University of Seville which allows the comparison between GENOM and one important reference tool for analog design, the FRIDGE optimizer [8].

6.6.1 FRIDGE Benchmark Circuit Tests

The benchmark circuit of reference is a novel single ended folded cascode OpAmp tested with FRIDGE synthesis tool [8], whose results are used to compare the performance and effectiveness of the final GENOM optimizer. This benchmark circuit includes all items necessary to the implementation and test, including the original netlist, testbenchs, device models, performance measures, constraints, range of variables and performance results obtained by the FRIDGE optimization tool. With this data, GENOM is able to test exactly in the same conditions as the FRIDGE tool. The schematic of the circuit is shown in Fig. 6.27 and testbench defined in Fig. 6.28.

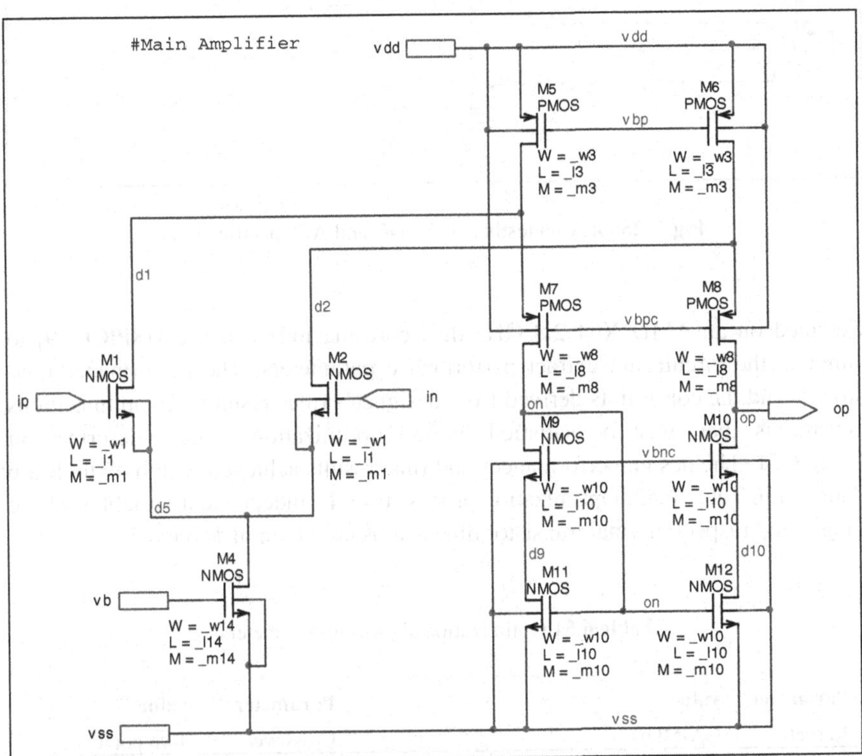

Fig. 6.27 Main Amplifier

6.6.2 Optimization Test with FRIDGE Ampop

Following the original FRIDGE approach, this experiment does not optimize the bias circuit, only the main circuit. The experiments were synthesized with the UMC 0.18um Regular Vt 1.8V Mixed Mode process Spice Model and were

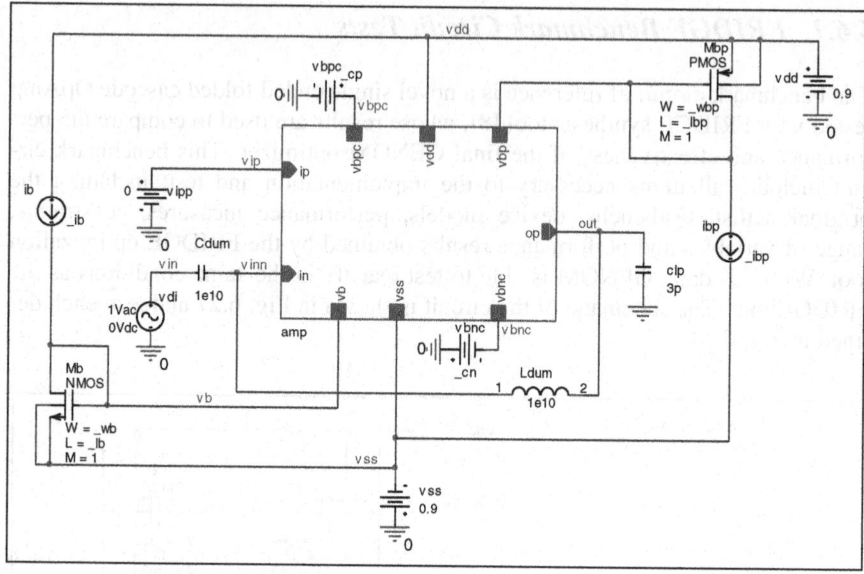

Fig. 6.28 OpAmp testbench for DC and AC specifications

executed on an AMD X64 2.8 GHz dual core machine and use HSPICE [9] to simulate the circuit and extract performance parameters. The performance constraints and the constraints derived from designer's rules result in 20 optimizations constraints that must be satisfied by the optimization process described in Table 6.51. The design performances and final results achieved with both tools are depicted in Table 6.52. Optimization process uses 15 independent variables whose ranges and respective final transistor dimensions are given in Table 6.53.

Table 6.51 Optimization algorithm parameters

Parameter	value	Parameter	value
Kernel	GA-MOD	Crossover	Two point
Strategy	Typical + Corner Optimization	Mutation	Dynamic
Sampling	LHS	Adaptive	No
Sort method	Priority to constraints then performance fitness	Elite	25% of population
Selection	Tournament by "feasibility"	Generations	150
Popsize	32	Search Space	4.716883e+53

Table 6.52 Design performance and final results

Target	FRIDGE	GENOM	GENOM Test specification
gbw > 1.20e+07	1.603e+07	**1.535e+07**	(gbw > 1.2e+07)
gain > 7.00e+01	7.000e+01	**7.061e+01**	+ (gain > 70.0)
pm > 5.50e+01	8.064e+01	**7.960e+01**	+ (verify_bound(pm,55,90))
sr > 1.00e+07	1.533e+07	**1.536e+07**	+ (sr > 1.0e+7)
dm2 > 1.20e+00	9.785e+00	**9.245e+00**	+ (check_bound(dm2, 1.2,1000))
dm4 > 1.20e+00	5.200e+00	**1.568e+00**	+ (check_bound(dm4, 1.2,1000))
dm5 > 1.20e+00	2.214e+00	**1.836e+00**	+ (check_bound(dm5, 1.2,1000))
dm7 > 1.20e+00	1.055e+01	**8.171e+00**	+ (check_bound(dm7, 1.2,1000))
dm9 > 1.20e+00	3.055e+00	**2.807e+00**	+ (check_bound(dm9, 1.2,1000))
dm11 > 1.20e+00	1.9594+00	**1.653e+00**	+ (check_bound(dm11,1.2,1000))
onm2 > 1.00e-01	1.004e-01	**1.098e-01**	+ (check_bound(onm2, 0.100,1000))
onm4 > 3.00e-02	3.023e-02	**3.240e-01**	+ (check_bound(onm4, 0.030,1000))
onm5 > 3.00e-02	5.662e-02	**9.866e-02**	+ (check_bound(onm5, 0.030,1000))
onm7 > 3.00e-02	4.255e-02	**8.761e-02**	+ (check_bound(onm7, 0.030,1000))
onm9 > 3.00e-02	4.919e-02	**3.802e-02**	+ (check_bound(onm9, 0.030,1000))
onm11 > 3.00e-02	1.782e-01	**2.451e-01**	+ (check_bound(onm11,0.030,1000))
osp > 5.00e-01	6.253e-01	**5.660e-01**	+ (check_bound(osp, 0.5, 1000))
osn < -5.00e-01	-5.022e-01	**-5.057e-01**	+ (check_bound(osn,1000, -0.5))
Area (min)	2.371e+01	**1.6873e+01**	+ (min(area, 0, 30))
Power (min)	2.333e-04	**2.446e-04**	+ (min(rmspow, 0, 0.001))
Cost value	-0.292589	**8.0704e-02**	---
Iter 1st/ (last) solution	---- / 2497	**1110/ (2464)**	---
Time (s) 1st/(last) sol.	n.a.	**25.08/(53.68)**	---

The main performance spec *gbw* stands for gainbandwidth, *gain* means the dc gain, *pm* is the phase margin, *sr* is the slew rate and the optimization goal is to minimize both the area (*Area*) and power dissipation (*power*). Both, the optimization goals and constraints used in the experiments were defined by the original benchmark circuit. The electrical constraints, as defined by the original benchmark circuit, are illustrated in HPSICE style in expression (6.3):

```
.meas ac dm2 param ='abs(lx3(x1.m1)/lv10(x1.m1))'
.meas ac onm2 param ='vgs(x1.m1) - vth(x1.m1)'
.meas ac osp param ='($cp + abs(vth(x1.m8)))'
.meas ac osn param ='($cn - abs(vth(x1.m10)))'
```
(6.3)

Table 6.53 Ranges and Final Transistor Dimensions

Optimization Var.	FRIDGE	GENOM
$cn = [-0.4,0];	$cn = -8.755479e-02	_cn = -4.490000e-02
$cp = [0.0,0.4];	$cp = 6.247103e-02	_cp = 1.000000e-03
$l1 = [0.18u,5u];	$l1 = 1.560000e-06	_l1 = 1.380000e-06
$l4 = [0.18u,5u];	$l4 = 4.700000e-07	_l4 = 1.940000e-06
$l5 = [0.18u,5u];	$l5 = 3.800000e-07	_l5 = 3.700000e-07
$l7 = [0.18u,5u];	$l7 = 7.600000e-07	_l7 = 9.100000e-07
$l9 = [0.18u,5u];	$l9 = 2.060000e-06	_l9 = 8.900000e-07
$l11 = [0.18u,5u];	$l11 = 6.000000e-07	_l11 = 2.190000e-06
$ib = log[30u,400u];	$ib = 4.842000e-05	_ib = 4.851000e-05
$w1 = log[0.24u,200u];	$w1 = 1.951000e-05	_w1 = 1.491000e-05
$w4 = log[0.24u,200u];	$w4 = 3.034000e-05	_w4 = 6.990000e-06
$w5 = log[0.24u,200u];	$w5 = 7.131000e-05	_w5 = 3.678000e-05
$w7 = log[0.24u,200u];	$w7 = 1.045300e-04	_w7 = 6.304000e-05
$w9 = log[0.24u,200u];	$w9 = 6.562000e-05	_w9 = 3.145000e-05
$w11 = log[0.24u,200u];	$w11 = 3.080000e-06	_w11 = 7.320000e-06

6.6.3 Comparison Results

Table 6.54 shows the GENOM and FRIDGE performance side by side and also depicts the GENOM run-time information in several optimizations points. In order to achieve a computing independent comparison between the tools, the following analysis is based, exclusively, on the number of evaluations "*nEval*" and the main goals, related to the minimization of power and area. Anyway, the time information was not provided with the actual benchmark circuit. GENOM achieved the first solution in 25s approx. using 1110 evaluations and reached a similar performance to FRIDGE in 1461 evaluations, corresponding to an efficiency increase of 41%. One of the best solutions improves simultaneously the power in 17% and 15% in the area as described in Table 6.54 with 2064 evaluations. The GENOM optimization was able to produce 183 new feasible solutions. Fig. 6.29 shows the gain magnitude and phase for typical mean process and 50° C conditions.

Table 6.54 GENOM benchmarks

Target	nEval	Power (min)	Area (min)	Time (s)
FRIDGE Final results	2497	2.333e-04	2.371e+01	---------
GENOM				
1st Feasible Solution	1110	4.0590e-04	2.9727e+01	25.08
GENOM similar to FRIDGE	1461	2.284e-04	2.377e+01	32.47
GENOM better than FRIDGE	2064	1.918e-04	2.009e+01	43.33
Final Results	2464	2.446e-04	1.6873e+01	53.68

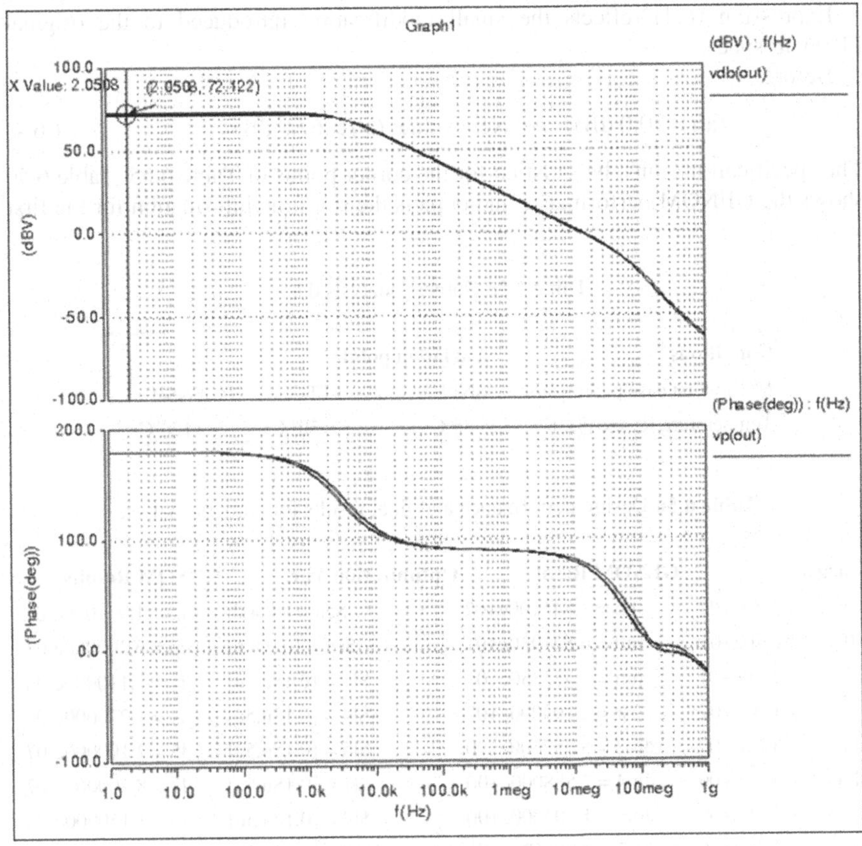

Fig. 6.29 Gain magnitude and phase for typical conditions

6.6.4 Corners Optimization with FRIDGE Circuit

Although there is no available benchmark information about the corner optimization for the FRIDGE benchmark circuit, the next experiment tests the GENOM performance for this type of optimization. However, there was the need to relax one specification, maintaining the others intact, in order to allow the corner optimization. This situation may occur when the performance specification is defined with a value that will not meet the worst-case corner point. The identification of this problematic specification was relatively easy to detect. First, it was verified that after several runs, the final solution always fulfils all constraints except one in a particular corner point. After identifying the problematic constraint, a new optimization was executed, assigning a high weight to this corner. However, the final solution did not improve, so this is probably the case where a specification was defined with a value that is not able to satisfy all corner points at the same time.

Expression (6.4) reflects the small modification introduced to the original FRIDGE specs.

Before:

$$(a0 > 70.0\ dB) \quad \rightarrow \text{ New value: } (a0 > 67.0\ dB) \tag{6.4}$$

The specifications must be satisfied for the corner points of Table 6.55. Table 6.56 shows the GENOM performance and depicts the run-time information for the first

Table 6.55 Corners analysis data

Conditions	Variation points		
MOS worst case parameters	SNFP	TT	FNSP
Temperature Range (° C)	-40° C	+50° C	+120°C

Table 6.56 Design performance and final results for corners analysis

Target		GENOM Results	Optimization Var.	GENOM Results
Gb	> 1.20e+07	gb = 1.845000e+07	$cn = [-0.4,0]	_cn = -1.971000e-01
a0	> 6.70e+01	a0 = 6.871930e+01	$cp = [0.0,0.4]	_cp = 6.300000e-03
pm	> 5.50e+01	pm = 7.435350e+01	$l1 = [0.18u,5u]	_l1 = 2.110000e-06
sr	> 1.00e+07	sr = 2.103000e+07	$l4 = [0.18u,5u]	_l4 = 1.270000e-06
dm2	> 1.20e+00	dm2 = 8.352600e+00	$l5 = [0.18u,5u]	_l5 = 4.100000e-07
dm4	> 1.20e+00	dm4 = 2.588000e+00	$l7 = [0.18u,5u]	_l7 = 8.100000e-07
dm5	> 1.20e+00	dm5 = 1.803000e+00	$l9 = [0.18u,5u]	_l9 = 1.150000e-06
dm7	> 1.20e+00	dm7 = 1.008620e+01	$l11 = [0.18u,5u]	_l11 = 2.420000e-06
dm9	> 1.20e+00	dm9 = 2.828500e+00	$ib = log[30u,400u]	_ib = 6.644000e-05
dm11	> 1.20e+00	dm11 = 1.695600e+00	$w1 = log[0.24u,200u]	_w1 = 2.496000e-05
onm2	> 1.00e-01	onm2 = 1.311000e-01	$w4 = log[0.24u,200u]	_w4 = 1.935000e-05
onm4	> 3.00e-02	onm4 = 1.695000e-01	$w5 = log[0.24u,200u]	_w5 = 4.813000e-05
onm5	> 3.00e-02	onm5 = 1.129000e-01	$w7 = log[0.24u,200u]	_w7 = 1.022000e-04
onm7	> 3.00e-02	onm7 = 6.762000e-02	$w9 = log[0.24u,200u]	_w9 = 4.983000e-05
onm9	> 3.00e-02	onm9 = 4.708000e-02	$w11 = log[0.24u,200u]	_w11 = 3.123000e-05
onm11	> 3.00e-02	onm11 = 1.362000e-01		
osp	> 5.00e-01	osp = 5.752000e-01		
osn	< -5.00e-01	osn = -6.185000e-01		
Areas	(min)	2.450920e+01		
Power	(min)	3.286000e-04		
Cost		1.145283e-01		
Iteration		20281		
Time (s)		411.18		

and final solution. This optimization produces 135 generations and executes 20281 electrical evaluations and creates 165 new solutions satisfying all design specs and functional constraints in all corners points.

Where, SNFP, TT and FNSP mean the Slow/Fast, Typical/Typical and Fast/Slow process, respectively.

Table 6.57 presents the final results for the present optimization problem.

Table 6.57 GENOM corner optimization

Performance Constr.	nEval	Power (min)	Area (min)	Time (s)
1st Solution in GENOM	9193	3.68E-004	3.31E+001	186.98
Final evaluation	20281	3.29E-004	2.45E+001	411.18

6.7 Conclusions

This chapter presented a set of experiments which test the GENOM's performance to design high-performance and novel circuit topologies. The above simulations have shown that the circuits designed by the GENOM tool conform to the synthesis objectives with efficiency and accuracy. Particularly, GENOM was able to achieve an efficiency increase of about 40% and a significant increase in performance when compared with one of the synthesis tool of reference.

The use of corners analysis and embedded designer rules methodology in every optimization run increases the value and trust in the final product, although the inclusion of corners analysis in the optimization scheme slows down the execution times considerably. This option produces a more robust design to parameter and process variations and in a certain way avoids the undesired circuits with high sensibility which causes big variations at the output in response to a small deviation in one of the parameters.

The great majority of the presented results are based on a 0.35μm CMOS technology because of the good availability of these models, although the GENOM tool has also been tested with success for a 0.18μm technology models in the telescopic and the FRIDGE OpAmp case studies. Since the technological process is independent from the optimization algorithm, virtually any technological process, including the more recent ones, can be supported by this tool.

With a proper configuration, the present optimization tool is able to synthesize a broad range of analog ICs beyond the class of circuits presented in this research.

References

[1] Wu, C.: Analog integrated circuits – lecture notes. IEE 6703, National Chiao Tung University (2006), http://www.cc.nctu.edu.tw (Accessed March 2009)

[2] Baker, R.J.: CMOS, circuit design, layout and simulation, 2nd edn. IEEE Press, John Wiley & Sons, Inc., (2005)

[3] Martin, K., Johns, D.: Analog integrated circuit design. John Wiley & Sons Inc., Chichester (1996)

[4] Milenova, B.L., Yarmus, J.S., Campos, M.M.: SVM in oracle database 10g: Removing the barriers to widespread adoption of support vector machines. In: Proc. 31st International Conference on Very Large Data Bases, pp. 1152–1163 (2005)

[5] Boardman, M., Trappenberg, T.: A heuristic for free parameter optimization with support vector machines. In: Proc. International Joint Conference on Neural Networks, pp. 610–617 (2006)

[6] Imbault, F., Lebart, K.: A stochastic optimization approach for parameter tuning of support vector machines. In: Proc. 17th International Conference on Pattern Recognition, vol. 4, pp. 597–600 (2004)

[7] Chang, C., Lin, C.: LIBSVM: A library for support vector machines (2001), `http://www.csie.ntu.edu.tw/~cjlin/libsvm` (Accessed March 2009)

[8] Medeiro, F., et al.: A Statistical optimization-based approach for automated sizing of analog cells. In: Proc. ACM/IEEE Int. Conf. Computer-Aided Design, pp. 594–597 (1994)

[9] Horta, N.C.: Analogue and mixed-signal systems topologies exploration using symbolic methods. In: Proc. Analog Integrated Circuits and Signal Processing, vol. 31(2), pp. 161–176 (2002)

7 Conclusions

Abstract. This last chapter presents the work conclusions and discusses the future work issues.

7.1 Conclusions

In this dissertation the application of evolutionary strategies to analog IC optimization problem has been discussed. It was developed a new approach to multi-objective and multi-constrained optimization technique for circuit sizing of analog circuits, which combines a robust optimization with corners and sensitivity analysis, machine learning and distributing processing capability. Particularly, a new hybrid optimization algorithm has been developed combined with a design methodology, which increases the efficiency on the analog circuit and system design cycle. This new algorithm combines an enhanced GA kernel with an automatic learning machine based on SVM model, which efficiently guides the selection operator of the GA algorithm avoiding time-consuming SPICE evaluations of non-promising solutions. The SVM model can be used as a feasibility or performance model. Whenever the model is built before optimization (offline) and the topology remains the same, it can be reused for other optimization runs with different performance requirements. Although the optimization tool is able to deal with equation based optimization, (as long as design equation has already been defined by an expert designer), the primarily decision is oriented to a simulation based approach, since it can be applied to all types of design circuits, producing more accurate results and providing an extended layer of analysis, concerning the robust design required in the industrial environment. Parameter variation effects due to manufacturing tolerances or environment conditions have also been included in the optimization loop implemented as a two step optimization methodology. The final solution results in a more robust approach with respect to variations and mismatches. Additionally, the undesired sensitivity effects are attenuated automatically by robust design.

The result of design methodology and optimization strategy is materialized in a tool, GENOM. The proposed design optimization tool represents an automated alternative to the traditional design flow, automating some steps of the design methodology. It covers some of the most time consuming tasks of the analog design process at the circuit or transistor level, like circuit sizing and design trade-offs identification. Like in many analog design environments, some time is spent in the optimization setup prior to synthesis runs. This includes the conformance test to the format of input files, configuration of optimization, definition of design and independent variables, definition of performances and respective measures, incorporation of technology models, corners, mismatches, designer rules and finally,

M.F.M. Barros et al.: Analog Circuits and Systems Optimization, SCI 294, pp. 187–189.
springerlink.com

the training of the learning model in case of optimization with offline model generation. All these tasks take advantage of the GUI interface developed in AIDA.

This computational tool allows a designer to examine regions of feasibility with differing uncertainty models available to approximate multi-objective problems like uniform distribution, latin hyper sampling and design of experiments. This tool also permits combining different algorithm approaches, like variations of standard operators and including several model approaches. A designer can quickly access promising design space regions by entering the historical database used to build the SVM model or consulting the database of non-dominated solutions where all the detailed information associated with the current problem is maintained. The graphical representation of the evolution process updated on the fly depends on the specifications provided by the designer. A summary of statistics in the form of post-processing text reports completes the feedback of the process. The information gained in one experiment was useful to the understanding of the overall problem. Further optimizations could be followed after the changing of some design or optimization parameters. Embodying this tool in a design platform or using it as a standalone application can lead to the increase of design efficiency and the improvement of the circuit performance, as it is demonstrated on several examples where the convergence to the desired performance criteria has been attained. The computation cost for several experiments have shown that circuits of moderate complexity can be synthesized in a reasonable amount of time using automatic learning models. This has been made possible by employing fast SVM models in the evolutionary cycle avoiding expensive simulation iterations. The synthesized designs have also been simulated and verified with HSPICE using the industry standard transistor models, such as the Alcatel and AMS. The simulations have shown that the circuits designed using GENOM conform to the synthesis specifications.

7.2 Future Work

In the domain of analog design automation, the research is always present and dynamic. There is yet, a definitely long way to end with the design gap between the improvement in manufacturing productivity and the progress in productivity achieved by CAD tools and design methodologies. Based on this work and involving the application field, some suggestions for future research are here provided:

General Topics

1. One of the major challenges related to this field of application is concerned with the development of a complete analog design automation environment involving the presented system with automatic topology selection and chip layout generation modules. The incorporation of layout information, for example, in optimization sizing process improves the robustness and reliability of the design solutions.

2. An alternative approach to the last item would be the integration of the present system within an industrial design environment, such as CADENCE framework. By developing the appropriated software interface modules, the presented optimizer tool can be incorporated as an external module to this commercial framework. The interface module and customized work environment may be implemented with the SKILL programming language.
3. Another area of potential research is related with topology generation. The present evolutionary computation technique with modeling technique can be used to implement an automatic search for new circuits and system topologies. The exploration methodology can be constrained to data structures from a specific design knowledge base or can be based on a random strategy aimed to explore new type of circuits.

Specific Topics

1. Develop an improved version of the GUI interface in order to reduce the setup effort to add a new circuit to database and improve the graphic information especially for online optimization.
2. Improve the GENOM Application Programming Interface (API) to allow an easy integration not only with AIDA design automation environment but with other CAD tools such as LAYGEN, etc.
3. Explore new methods of parallel processing offered by the "Open MPI", the new release of the "open source high performance computing" message passing library, implemented in GENOM. The actual implementation architecture uses a master-slave parallel architecture but other architectures could be investigated.
4. Use the information achieved by the design space exploration and respective trade-offs for all circuits from the library in order to perform an accurate topology selection.
5. The experiments developed in this thesis were driven essentially for the design of continuous-time class of amplifiers. An extension of this tool should support other types of circuits.

Appendixes

Appendix A. Terminology

Table A.1 Control parameters

Term	Definition
Design Objectives or Design Goals	Corresponds to the minimization or maximization of one or several objectives, for example: *min,max (power, area,...)*
Performance Specs	A set of values which indicate levels of performance in order to ensure a certain functionality of the circuit. They are design objectives usually defined in the form of inequality constraints, such as, *gain > 70 dB; gbw > 100MHz, etc.*
Design constraints	**(i) Parameter constraints** – Formed by device sizes range, e.g., W= [min,max] = [0.18, 100]μm, L=[0.18, 10]μm, etc. **(ii) Functional constraints** – Corresponds to the requirements of some basic electric design requirements, e.g., saturation of certain transistors, etc., in order to ensure the correct circuit operation, e.g., the overdrive voltage, $(V_{GS} - V_{Th})$ is defined in [50-200] mV. **(iii) Performance constraints** – Design objectives usually in the form of inequality constraints, such as, e.g., *gain >70 dB*; *gbw > 100 MHz*.
Optimization parameter	A set of independent decision variables of an optimization problem. In circuit sizing problems usually they correspond to the design parameter constraints, e.g., the widths (w) and lengths (l) of transistors.
Fitness function	A fitness function is a particular type of objective function that quantifies the optimality of a solution. Used to rank a particular solution.
Cost function	A particular type of fitness function which assigns a better rank to solutions with lower fitness.
Merit function	A particular type of fitness function which assigns a better rank to solutions with higher fitness.
Optimal solution	A solution to an optimization problem which minimizes (or maximizes) the objective function.
Design Space	**(a) Design Space** (DS) – A multidimensional space delimited by the ranges of parameter constraints. **(b) Goal Space** - A subset of the multidimensional DS that satisfies all design objectives and performance constraints. **(c) Functional Space** – A subset of the multidimensional DS formed by the interception of all functional constraints. In this thesis, it is defined as feasibility space. **(d) Solution Space** – A portion of the design space that satisfies simultaneously the performance and feasibility region.

Table A.1 (*continued*)

Term	Definition
Performance Space	(e) **Performance Space** (PS) – It is the space of all possible performance values based on the evaluation of all points from the design space. The mapping between design parameters (D) and the performance space, D→P(D), is usually done by circuit simulation with spice-like analog simulators. (f) **Performance Region** – It is the region in the PS achieved by a subset of all individuals in the DS that satisfies all the performance inequality constraints. (g) **Feasibility Region** – It is the region in the PS achieved by a subset of points in the DS that satisfies all the functional constraints. (h) **Feasible Region** – It is the region in the PS achieved by a set of points in the DS that satisfies both the performance constraints, as well as, the functional constraints. It is the region of all possible solutions of an optimization problem. (i) **Infeasible Region** – Set of points outside the feasible region.

Fig. A.1 is a 2D sketch for the conceptual terms introduced above. The multidimensional axis *d1*, *d2* represents the parameter constraints, *Ws*, *Ls* and *Ms* of transistors. The multidimensional axis *p1*, *p2* represents the design specs, e.g., *gain*, *gbw*, *power*, etc.

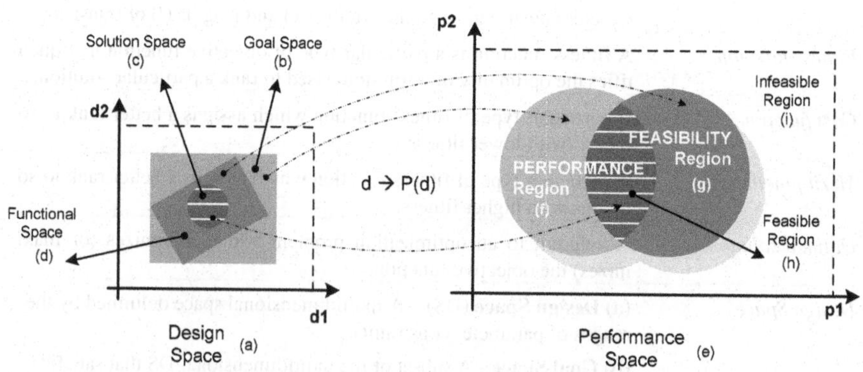

Fig. A.1 Conceptual view of design spaces adopt in the terminology

Appendix B. General Purpose Optimization Techniques

B.1 Random Search Methods

In essence, they simply consist of selecting randomly potential solutions and evaluating them. They do not use any heuristics (or meta-heuristics) to guide the next potential solution, so the search is very slow. The best solution over a number of samples is the result of "pure" random search, p.e., the Monte Carlo (MC) method. In spite of being considered the weakest of all optimization methods, random search methods have some visibility once they are often used as a reference tool. One of the first improvements to random search is given by the simple Hill Climber algorithm (Fig. B.1) and is applied to non-linear unconstrained problems.

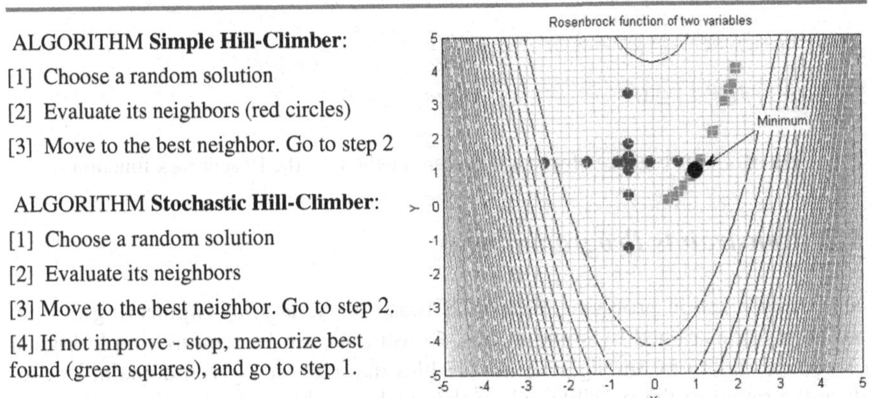

ALGORITHM Simple Hill-Climber:
[1] Choose a random solution
[2] Evaluate its neighbors (red circles)
[3] Move to the best neighbor. Go to step 2
ALGORITHM Stochastic Hill-Climber:
[1] Choose a random solution
[2] Evaluate its neighbors
[3] Move to the best neighbor. Go to step 2.
[4] If not improve - stop, memorize best found (green squares), and go to step 1.
Rosenbrock function of two variables
Minimum
X
Y

Fig. B.1 The Hill-climber Algorithm

B.2 Unconstrained Gradient-Based Methods

Gradient based methods belong to the class of unconstrained non-linear optimization algorithms (Fig. B.2) which apply the concept of successive search within the optimization space, based on the information of gradient or derivative function [1]. To be efficient the cost function should be unimodal (single local optimum), continuous and differentiable. The iterative process perturbs the current vector position to obtain the next value $\vec{X}_{k+1}$. Normally, this iteration is given by $\vec{X}_{k+1} = \vec{X}_k + \lambda_k \vec{d}_k$ where $\vec{d}_k$, indicates the direction of the next move and the step size λ_k controls the evolution and the precision of the solution. *Golden section*, cubic interpolation and *Fibonacci* techniques can be used to determine the value of λ_k and the direction $\vec{d}_k$ can be determined by the *Newton method* as well as other methods. Each point in the generated sequence has a lower cost than its predecessor. The weakness of this method is that line minimization may be expensive and convergence can be too slow for ill-conditioned problems. Also, when the derivative (or an approximation to the derivative) can-not be determined these methods cannot be used.

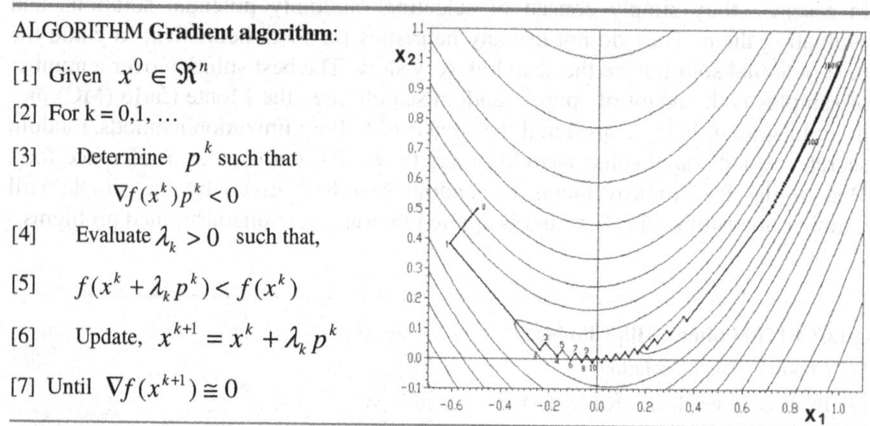

ALGORITHM **Gradient algorithm:**
[1] Given $x^0 \in \Re^n$
[2] For k = 0,1, …
[3] Determine p^k such that $\nabla f(x^k)p^k < 0$
[4] Evaluate $\lambda_k > 0$ such that,
[5] $f(x^k + \lambda_k p^k) < f(x^k)$
[6] Update, $x^{k+1} = x^k + \lambda_k p^k$
[7] Until $\nabla f(x^{k+1}) \cong 0$
X2
X1
1.1
1
0.9
0.8
0.7
0.6
0.5
0.4
0.3
0.2
0.1
0.0
-0.1
-0.6
-0.4
-0.2
0.0
0.2
0.4
0.6
0.8
1.0

Fig. B.2 Convergence of steepest descent method for the Rosenbrock function

B.3 Constraints Programming

Many optimization problems, especially those in the area of engineering design, are highly constrained by some means. Constraints can be seen as simple logical or numeric relations among several variables that restrict a given domain, i.e., reduce the range of the possible values that each variable can take. Constraints can usually be expressed in terms of function in-equalities, strict inequalities or equality constraints as exemplified in Fig. B.3. There are several ways of dealing with constraints. The classes of problems modeled by integer linear programming techniques are usually solved by two mature tools like the simplex algorithm and the Constraint Satisfaction Problem (CSP) techniques [2]. The most common approach to manage infeasible solutions uses the concept of penalty functions which transform the original constraint problem in an artificial unconstrained optimization problem. This alternative penalizes the solutions that are near or violate the constraints boundaries with an amount proportional to constraint violation. In this way the constrained problem can be solved using a sequence of unconstrained optimizations, which in the limit is expected to converge to the solution of constrained problem. This approach is generally associated with fitness assignment in some global optimization algorithms like evolutionary algorithms. A comprehensive survey of the most popular constraint handling techniques used for EAs can be found in [3]-[5].

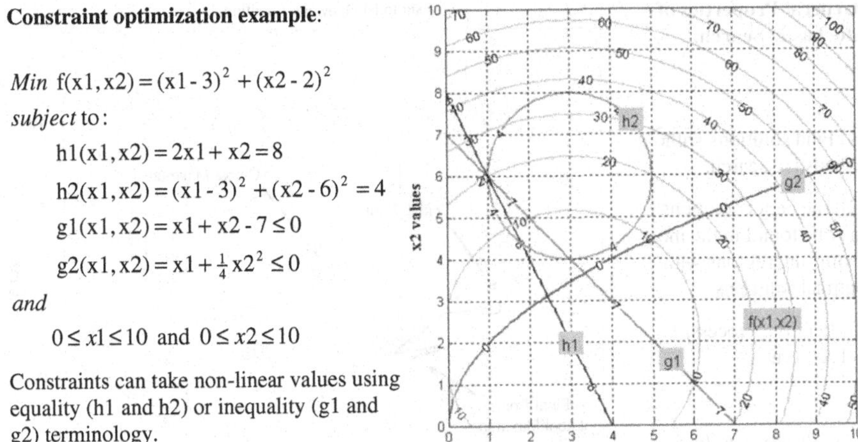

Constraint optimization example:
Min f(x1,x2) = (x1 - 3)² + (x2 - 2)²
subject to:
h1(x1,x2) = 2x1 + x2 = 8
h2(x1,x2) = (x1 - 3)² + (x2 - 6)² = 4
g1(x1,x2) = x1 + x2 - 7 ≤ 0
g2(x1,x2) = x1 + ¼x2² ≤ 0
and
0 ≤ x1 ≤ 10 and 0 ≤ x2 ≤ 10
Constraints can take non-linear values using equality (h1 and h2) or inequality (g1 and g2) terminology.
h1
h2
g1
g2
f(x1,x2)
x1 values
x2 values

Fig. B.3 Constraint optimization problem

Briefly, these methods increase the efficiency of the search using the constraints to prune the search space. Constraint-based systems derived from OR field, normally use a declarative way of programming which makes easier the modeling of complex problems, their modification and maintenance.

B.4 Direct Stochastic Methods

This stochastic programming class encloses a broad range of distinct algorithms that do not require a continuous, a convex or differentiable cost function. Therefore it does not need to derive or compute gradients or take care of discontinuities. Stochastic algorithms outperform one of the major drawbacks of simple deterministic algorithms, i.e. they are particularly effective when the goal is to find an approximate global optimum for multimodal functions (Fig. B.4). They own some other intrinsic advantages, allowing simple implementations and flexible formalization of the problem, handling multimodal and noise functions, solving discrete and combinatorial problems, as well as, being in some cases well suitable for parallel computing [6]-[7].

Stochastic search algorithms is an umbrella set of methods that include the Nelder-Mead simplex-based methods [8], the simulated annealing (SA) [9]-[10], Tabu search (TS) and evolutionary algorithms where the Genetic Algorithms appear as one of the most notorious in this class. The first one inherited its name from the n+1 geometric figure in n-dimensional space called a simplex; the second is based on the physical process of annealing the materials; the third one applies the concept of memory maintaining a "tabu list" of solutions already vi-sited, while the last ones emulate some kind of nature's evolutionary behavior. They differ in some implementation details but all share a common approach, the search for the optimal value which follows some probabilistic rules in order to make the new generation of solutions better than the previous one.

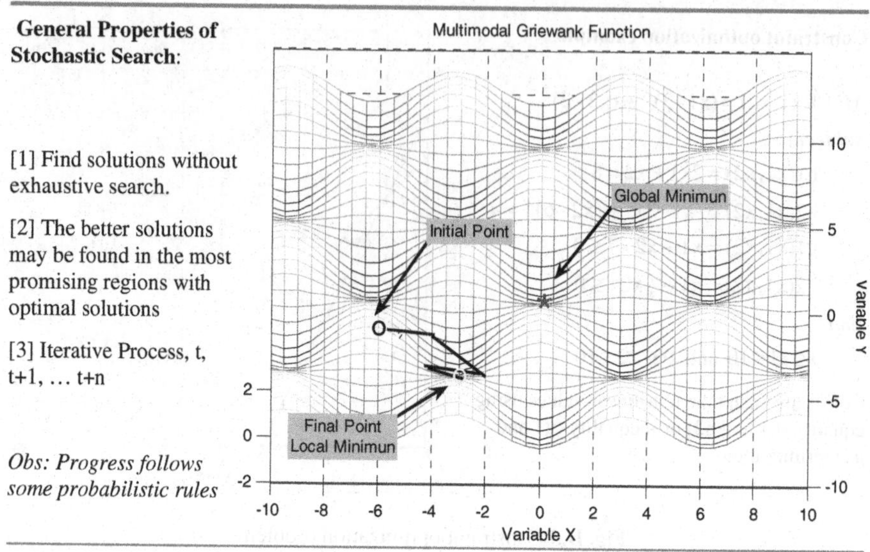

General Properties of Stochastic Search:
[1] Find solutions without exhaustive search.
[2] The better solutions may be found in the most promising regions with optimal solutions
[3] Iterative Process, t, t+1, ... t+n
Obs: Progress follows some probabilistic rules
Multimodal Griewank Function
Global Minimun
Initial Point
Final Point
Local Minimun
Variable X
Variable Y
-10
-8
-6
-4
-2
0
2
4
6
8
10
5
-5

Fig. B.4 General properties of Stochastic Search algorithms

The simulated annealing (SA) approach [9], for example, is a numerical optimization technique based on the principles of the cooling process of some materials. Unlike EAs, the progress in the search space is supported by a single individual. The algorithm starts from a valid solution and randomly generates a new state (point in the search space) which is immediately evaluated, as described in Fig. B.5. If a better solution is found (*New_Cost- Current_Cost<=0*), the new solution has lower cost and so it is immediately accepted (k_1 point), if not (k_2 point), that solution can only be accepted with some probability that depends on the environment temperature *T*. In the beginning of the process, *T* starts with high virtual temperature and progressively slows down its values. The interesting effect produced by this changing in temperature is to allow a better exploration of the search space in the beginning of the process and less exploration, i.e., better exploitation at the end of the process. In other words, the probability of accepting a worse state, given by the expression $\mathrm{Prob} = \exp(-\frac{\mathrm{New\_Cost} - \mathrm{Current\_Cost}}{\mathrm{Temp}})$, is high at the beginning and decreases as the temperature decreases. This phenomenon known as the Metropolis criterion is expressed by computation code of Fig. B.5.

Whereas some of the algorithms like SA and TS are guided solely by random rules with no sense of the appropriate direction or size of step to take, other methods like GA and ES correct this conduct by means of heuristic operators. Because of their probabilistic nature, the convergence to the global optima usually requires many iterations. But with the recent progress in computer systems and distributing computing techniques, the stochastic methods have gained great popularity. These myriad of characteristics make them appropriate for a wide variety of optimization

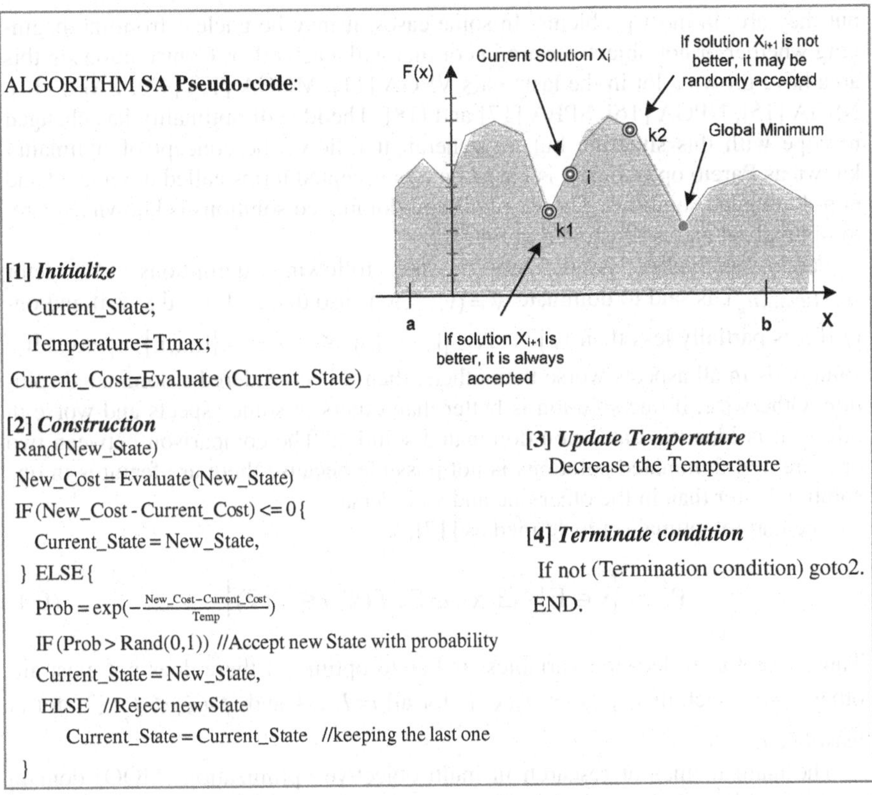
ALGORITHM SA Pseudo-code:
F(x)
Current Solution Xi
If solution Xi+1 is not better, it may be randomly accepted
k2
Global Minimum
i
k1
a
b
X
If solution Xi+1 is better, it is always accepted
[1] Initialize
Current_State;
Temperature=Tmax;
Current_Cost=Evaluate (Current_State)
[2] Construction
Rand(New_State)
New_Cost = Evaluate(New_State)
IF (New_Cost - Current_Cost) <= 0 {
Current_State = New_State,
} ELSE {
Prob = exp(- New_Cost-Current_Cost / Temp)
IF (Prob > Rand(0,1)) //Accept new State with probability
Current_State = New_State,
ELSE //Reject new State
Current_State = Current_State //keeping the last one
}
[3] Update Temperature
Decrease the Temperature
[4] Terminate condition
If not (Termination condition) goto2.
END.

Fig. B.5 The basics of Simulated Annealing algorithm

problems, covering a broad field of applications, including the analog design problem. A trade-off between a large spectrum of applications and performance efficiency is explained by the free lunch theorem described in section 3.1.3.

B.5 Multiple Objectives

In engineering and control applications it is common to deal with problems requiring the optimization of more than one objective function instead of just one. A typical example is car engine design, where the task may be to maximize the performance while minimizing the fuel consumption. A multi-objective optimization problem (MOO) usually involves a number of conflicting objectives that have to be handled simultaneously. It is rarely the case where a single point simultaneously optimizes all the objective functions of a multi-objective problem. Therefore, the solution of this type of problem is supported by illustrative trade-offs of objec-tive functions rather than in a single solution allowing a final human decision among the solutions. The objectives do not necessarily have to be conflicting, but they are, in most problems. In some cases, it may be unclear from the beginning whether or not objectives are in conflict with each other. Contributions in this area have grown a lot in the last years VEGA [11], MOGA [12]-[13], PAES [14], NSGA [15], NPGA [16], SPEA [17] and [18]. The idea of optimality has changed to cope with this situation but, in general, it follows the concept of optimality known as Pareto optimum. It is based on two accepted terms called dominated and non-dominated solutions. The set of all non-dominated solutions is known as Pareto-optimal set and is illustrated in Fig. B.6.

Pareto optimality is defined by the following definitions. A vector $\vec{u} = (u_1,\ldots,u_k)$ is said to dominate $\vec{v} = (v_1,\ldots,v_k)$ also denoted by $\vec{u} \le \vec{v}$ if and only if u is partially less than v, i.e. $\forall \mathrm{i} \in \{1,\ldots,\mathrm{k}\}, \mathrm{u_i} \le \mathrm{v_i} \wedge \exists \mathrm{i} \in \{1,\ldots,\mathrm{k}\}: \mathrm{u_i} < \mathrm{v_i}$. If a solution is in all aspects worse than others, then it is considered a dominated solution. Otherwise, if one solution is better than others in some aspects and worse in others, it is identified as a non-dominated solution. The comparison between two or more non-dominated solutions is not possible because there are features in one solution better than in the other one and vice-versa.

The Pareto optimal set is defined as [17]:

$$\mathrm{P}^* := \{\vec{\mathrm{x}} \in \mathrm{F} \mid \neg\exists\ \vec{\mathrm{x}}^* \in \mathrm{F} : f(\vec{\mathrm{x}}^*) \le f(\vec{\mathrm{x}})\} \tag{B.1}$$

Thus, a vector of decision variables is Pareto optimal if there does not exist another $\vec{\mathrm{x}} \in \mathrm{F}$ such that $f_i(\vec{\mathrm{x}}) \le f_i(\vec{\mathrm{x}}^*)$ for all $i=1,\ldots,k$ and $f_j(\vec{\mathrm{x}}) \le f_j(\vec{\mathrm{x}}^*)$ for at least one j.

The main themes of research in multi-objective optimization (MOO) domain are focused in techniques for handling constraints, maintaining diversity of the solutions, hybridization with other local search methods and archiving for storing non-dominated vectors.

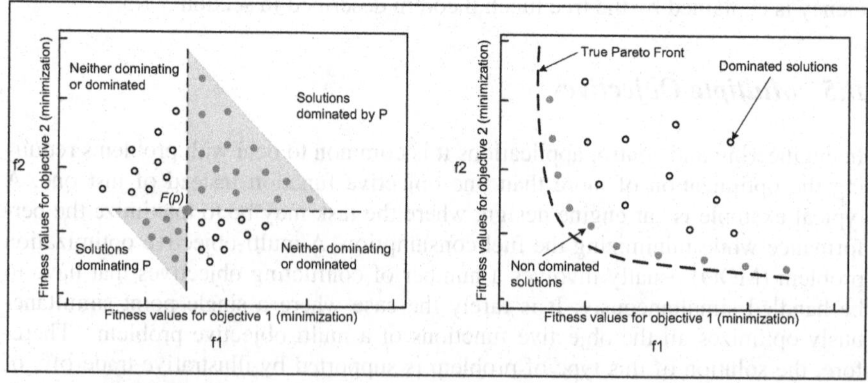

Fig. B.6 Dominance, non-dominance and Pareto Front in MOO problems

Appendix C. The Basic Decisions of Standard GA Algorithms

C.1 Standard GA Kernel Optimization

C.1.1 Evolutionary Kernel Framework

The major task of EC techniques and in GA in particular is to compute artificial models simulating an evolutionary process. They differ from more traditional search algorithms in that they work with a population of candidate solutions that will evolve progressively towards a certain goal. Meanwhile the algorithm iteratively applies probabilistic transformations to the population and uses a selection scheme to obtain an improved population. This goal is to find the best possible approximate solution of a given complex optimization and design problem. The artificial models mimic natural evolution in a simplified way. The three main mechanisms used to drive evolution forward are depicted in Fig. C.1.

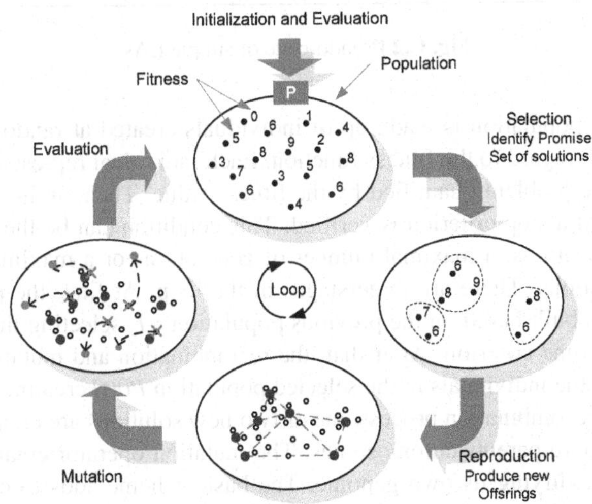

Fig. C.1 Common evolution cycle

Fig. C.1 illustrates a typical iterative cycle of evolutionary algorithms and the three main mechanisms used to drive evolution forward, namely, reproduction, mutation, and selection. The fundamentals of EAs are based on the existence of a population of individuals that will change dynamically in each generation (each loop cycle) through the influence of operators mimicking the biology cycle of life.

As a matter of fact all the EAs terminology was inherited from biology life. In spite of this, an individual is a compound structure forming a chromosome, fitness, and possibly a number of other attributes. The chromosome encapsulates a sequence of genes representing a solution of the problem. The chromosome defines the interface between the problem and the optimization algorithm. The fitness function is a measure of the quality of an individual represented by the chromosome.

Fig. C.2 illustrates the classic structure of a simple evolutionary algorithm introduced by Holland [19] and known as the basic genetic algorithm flow.

```
EA Main Procedure:
t=0
initialize (Pop(0))
evaluate (Pop(0))
while (! (Termination condition)) {
    t=t+1;
    P'(t)  = select (Pop(t-1))
    P''(t) = Recombine (P'(t))
    P(t)   = Mutate (P''(t))
    Evaluate (P(t))
}
```

Fig. C.2 Pseudo-code of simple EAs

In EA, the population is made up of individuals created at random, which are evaluated with regard to the fitness function. Each individual represents a potential solution of the problem quantified by the fitness value. Then, an iterative process is applied until a stop criterion is verified. This condition can be the achievement of the desired fitness, a maximal number of generations or a maximal number of fitness evaluations. There are three stages in the loop. At first, the population at generation *t* is built based on the previous population *t-1*, selecting the fittest individuals with some criterion. After that, the recombination and mutation operators are applied to the individuals in the selected population $P0(t)$ creating a new population. In the recombination process one or two new solutions are created, crossing over two or more parents chromosomes. The mutation operator creates a new individual by modifying its own genome. The basic scheme adds to chromosomes some type of stochastic noise. At last, in the final stage, the evaluation of the new population is carried out and the whole process is repeated.

C.1.2 Algorithm Design Parameters

In order to guide a population of candidate's solution towards an optimum, many decisions have to be taken, which have a deep influence on the effectiveness and efficiency (see definition in Appendix C.8) of the algorithm. The basic decisions of standard GA algorithms include the choice of the most suitable structure and genetic representation, the selection and replacement strategy, the crossover and mutation parameters and other algorithm control parameters. The operation of simple GAs is managed by a set of control parameters that have great impact on the performance of the algorithm. These control parameters include, the probability of mutation and crossover, the tournament size of selection or the population size (number of individuals in the population), which will be explained in the Table C.1.

Table C.1 Control parameters

Control Parameter	Impact
The population size [20]	The population size is the number of individual organisms in a population participating in the evolutionary process and it has great impact on the computation time per iteration; if the population size is too large, the algorithm tends to take longer time to converge, but if the population size is too small, the GA is in risk of premature convergence because there may not be enough diversity in the population to let the GA escape from local optima. Common values observed in literature adopt values between 30 and 200.
The crossover rate [20]-[22]	The crossover rate defines the frequency of the crossover operation which enables the evolutionary process to move towards the most promising regions of solution space. The crossover probability *px* has the function of controlling the rate/frequency at which individuals are submitted to crossover. If the value of *px* is high, the new solutions will be quickly applied into the population, but if the value is too high, individuals may be disrupted faster than selection can exploit them. For this reason the *px* usually takes the values from 0.5 to 1.0.
The mutation rate [20]-[22]	The mutation rate is expressed by a probability and has influence on the diversity in the population. Both a high and a low mutation rate have disadvantages: a high rate causes high diversity in the population, transforming the GA into a random search algorithm whereas a low mutation rate makes it hard to achieve a global optimum solution because convergence may occur too early, producing premature convergence to a local optimum. The typical values for mutation rate are chosen in the range 0.001 to 0.05.

The estimation parameters presented in Table C.1 result from study cases found in literature, normally applying standard GA settings or configurations. There is also no magic number or deterministic formula concerning the optimal settings or optimal control of these parameters over the time, or when changing the dimension of search space, the length or coding representation. Under different configurations, e.g., a different problem codification, these parameters can achieve higher values. The study of ideal control parameters configuration for a given problem is a time consuming task mostly based on experiments. This approach, however, has several disadvantages. As the control parameter behaviors are not independent, systematic trials executed for all possible combinations are almost impossible.

Procedures to control parameters consume a lot of computation time and the achieved control parameters may not be necessarily the optimal values.

Tuning these parameters by using a direct control mechanism before the running of the algorithm is a typical practice applied in genetic algorithms and their derivates but it is not efficient because it is known that no generally valid best parameter value exists. The use of adaptive techniques inspired from ES community is one alternative to get around with optimal settings of parameters. Instead of using rigid parameters that do not change during the evolutionary process, the idea is to control them during the run. Several techniques inherited from ES community are applied to change the mutation step size control such as, the *1/5th-success rule*, *cumulative path length control* and *self-adaptation* [21]. From these type of techniques it is verified a frequent supremacy of mechanisms depending on the distance to the optimum.

C.1.3 Single Optimization GA Example

In this section, the Genetic Algorithm will be applied to simple optimization problems. The numerical examples of constrained optimization problem are given in Table C.2 as follows:

Table C.2 Testbench functions for the GA optimization example

Minimization of Function 1	$F(x,y) = -x^2 - 2y^2 + \cos(3\pi x) + 0.3\cos(4\pi y) + 4$ $Constraints: -1 \le x \le 1; -1 \le y \le 1$
Minimization of Function 2	$Z1 = \sqrt{1 + x^2 + y^2};\quad Z2 = \sqrt{1 + (x-1)^2 + (y+1)^2}$ $Z3 = \sqrt{1 + (x+1)^2 + (y+1)^2}\quad Z4 = \sqrt{1 + (x-1)^2 + (y-1)^2}$ $Z5 = \sqrt{1 + (x+1)^2 + (y-1)^2}$ $F(x,y) = 5 - \sin(\frac{4*Z1}{Z1}) - \sin(\frac{2.5*Z2}{Z2}) - \sin(\frac{3*Z3}{Z3}) - \sin(\frac{2*Z4}{Z4}) - \sin(\frac{4*Z5}{Z5})$ $Constraints: -5 \le x \le 5; -5 \le y \le 5$

A three-dimensional plot of objective functions is shown in Fig. C.3 (a). For these continuous functions, the chromosome is encoded as a vector of two real values *x* and *y*. When a new population is created, the next step is to calculate the fitness value of each member in the population. The evaluation of the fitness for each chromosome is performed by *Eval(F(x,y))*. After the evaluation of the last chromosome a new population is created. The two most fitted individuals are reproduced directly in the next population while the remaining ones will be submitted to the standard process variations, crossover and mutations. The chromosomes selected to crossover will be chosen according to the roulette wheel strategy. The roulette wheel cumulative probability for each individual is calculated with expression C.1:

$$Q_i = \sum_{k=0}^{i} P_k \qquad (C.1)$$

Where, the selection probability Pi and the total fitness expression are given by:

$$P_i = \frac{Eval(f(x,y))}{F\_total} \;;\; F\_total = \sum_{k=0}^{iPop} Eval(f(x,y)) \qquad (C.2)$$

Table C.3 gives the configuration parameters of this single optimization problem.

Table C.3 Optimization control parameter configuration

Algorithm	Name	Function 1	Function 2
Initial Population / Population Size	iPOP/Pop	64 /32	32/16
Initial Sampling Method	sAMP	random	random
Elite population (survive to next gen.)	Elite	2	2
Selection type	sType	Roulette wheel	Roulette wheel
Crossover rate / type	cRate/cXover	50% / One point	50% / One point
Mutation rate / type	mRate/mtype	5% Fixed/random	5% Fixed/random
Number of generations	nGEN	10	10
Independent Variables	iVar	x, y	x, y
Kernel Type	Kernel	GA	GA

The results illustrated in Fig. C.3 b) and c) plots a two dimension view and the contour plot of the function under test with the initial population locations denoted by circles and the final solutions with red stars. Fig. C.4 shows the evolution curve of the best and average fitness across 64 generations for 10 runs of the algorithm. In a typical application the best curve presents a monotonically decreasing shape with respect to generation numbers.

At the end of each generation the fitness of the best individual is expected to improve whereas there is a tendency to stagnate to the end of the run (see Fig. C.4). The *stagnation* may be the consequence of several events. The more optimistic is the successful discovery of the global solution of the problem. In that case the algorithm cannot evolve any further while stop condition, perhaps the maximum number of iterations have not exhausted yet. However, the discovery of the global optimum does not always happen.

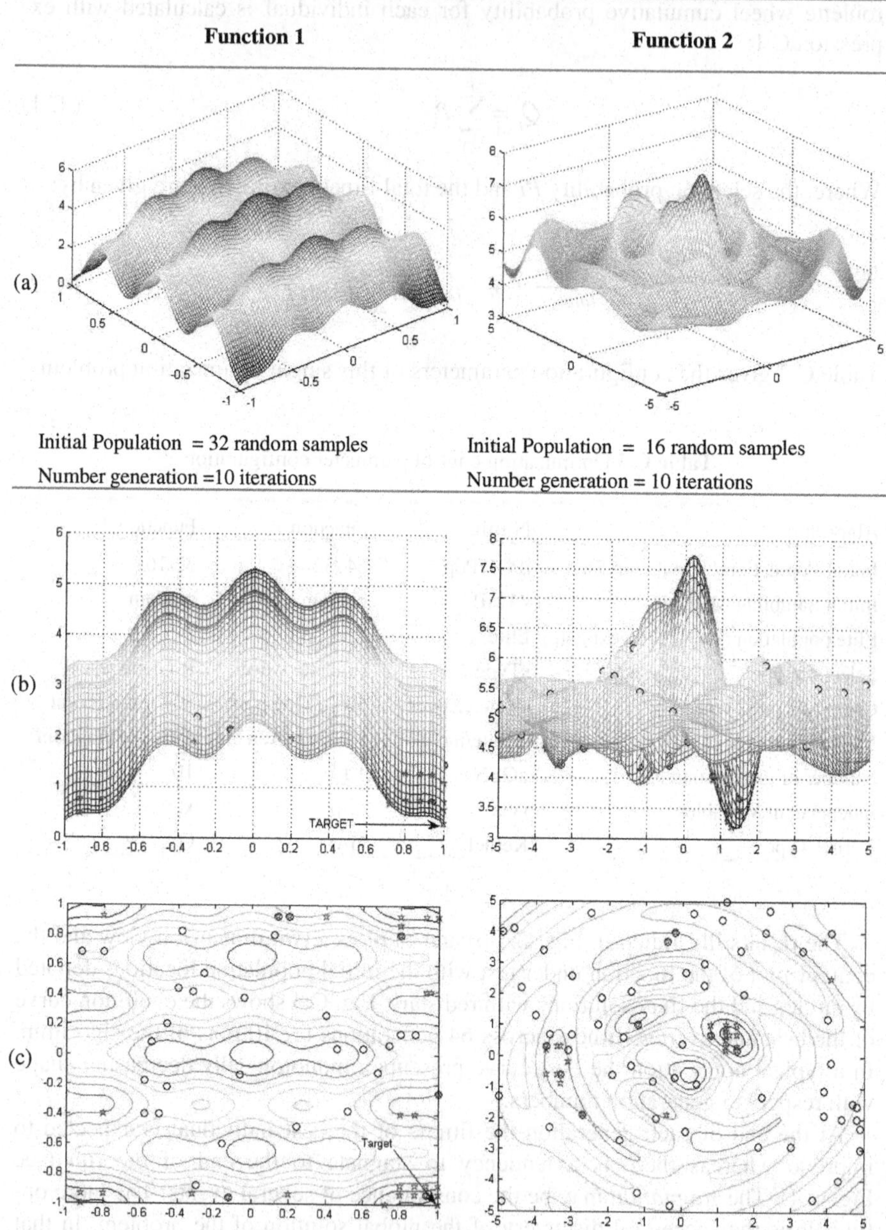

Fig. C.3 3D, 2D view and the contour plot of the function under test

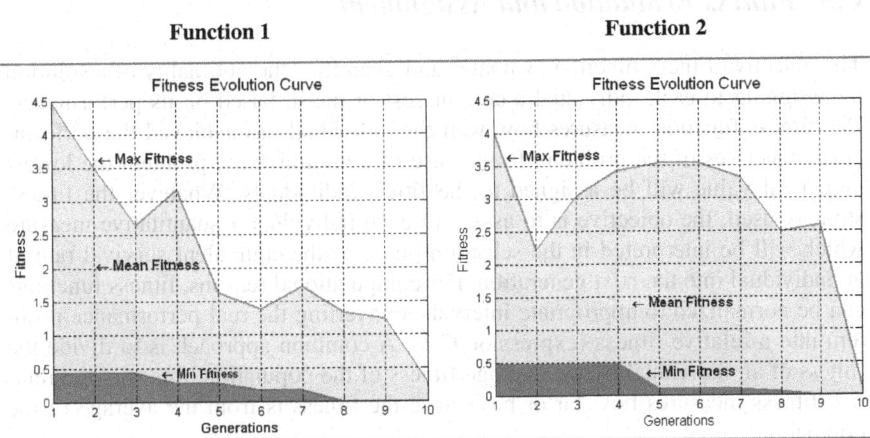

Fig. C.4 Illustration of the evolutionary process

Stagnation may occur because of the influence of a local optimum, an insufficient number of iterations cycles or wrong parameterization and/or bad choice of the search methods. These are the main common issues in EAs algorithms, as well as, in other iterative search methods.

C.2 Representation and Encoding

GAs are population-based searching algorithms whose individuals are represented by chromosomes with several genes encoded in binary or real coded form. The binary method used by the classical GA encoding system has some weak points, whenever it is used in multidimensional, high-resolution numerical problems [23]. The real-coded representation has more advantages as it is faster and more accurate in solving optimization problems whose parameters are represented in continuous domain.

Besides, the real-coded representation allows the creation of more sophisticated operators, thus it is the representation adopted in this thesis. Whichever the type of representation used, the chromosomes are usually implemented in the form of vector lists of attributes where each attribute, known as gene, is a representation of one optimization variable. Fig. C.5 illustrates the chromosome structure for some objective functions.

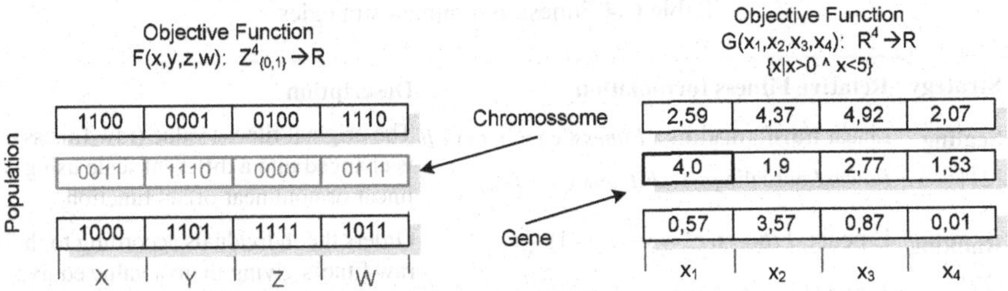

Fig. C.5 GAs basic structures

C.3 Fitness Evaluation and Assignment

The objective fitness function evaluates and quantifies the optimality of a solution by assigning to each individual a certain cost or merit, based on its performance. The fitness function measures how well the individual has achieved the performance objectives of the problem. In the case of a minimization problem the lowest numerical value will be assigned to the fittest individuals. Whatever the fitness strategy used, the objective is to assign to each individual a quantitative measure which will be interpreted in the selection phase as the equivalent survival rate of an individual into the next generation. For computational reasons, fitness functions can be normalized to appropriate intervals, converting the real performance problem into a relative fitness (expression C.3). A common approach is to divide the fitness of an individual by the average fitness of the population, this way the relative fitness measures how far or how close the fitness is from the average of the population.

$$Fitness(x_i) = g(f(x_i)) = \frac{f(x_i)}{\sum_{i=1}^{Nind} f(x_i)} \tag{C.3}$$

There are many methods to evaluate fitness and assign a real number to each chromosome. The fitness assignment strategies can be summarized in two essential types: Scaling Fitness and Ranking. Table C.4 presents two methods for transforming the objective function in a relative fitness, where N is the population size, r represents the rank and s the relative fitness for best individual such that, $1<s<2$.

When scaling, some precautions should be taken against the formation of very large relative fitness in some stronger individuals to get around with early premature convergence. In the same way, if there is not a clear differentiation between the performance of the best individuals and the rest of the population, the search will not be worthwhile. Both situations are not effective so the relative fitness scheme assigned by scaling should be carefully chosen. Rank-based fitness, on the other hand, is less sensitive to these unwanted scale effects since the appearance of super individuals responsible for early premature convergence is weakened because the best individual in the population is always assigned the same fitness, and in a population of similar performance values, the best one is still preferred to the rest.

Table C.4 Fitness assignment strategies

Strategy	Relative Fitness formulation	Description
Scaling [21]	Linear transformations: $Fitness(x_i) = af(x) + b$ Power law scaling: $Fitness(x_i) = f(x)^k$	The original fitness value (raw fitness) is changed into a different scale using linear or nonlinear offset function
Ranking [21]	Linear: $Fitness(r) = s - (s-1).\frac{2r}{N-1}$ Power law ranking: $Fitness(r) = \rho^r.s$	Orders the individuals according to their raw fitness giving them a value equivalent to their position in the ranking.

C.4 Initial Population

In evolutionary algorithms the initialization of the population specifies the starting points of the search. Traditionally, the initial population is created randomly but several other initialization techniques can also be adopted. The next paragraphs introduce two sampling techniques that will be used in this thesis, *design of experiments* (*DoE*) [24]-[25] and *latin hyper sampling* (*LHS*) [26].

DoE is a statistical experimental design methodology used to perform multivariate design experiments in order to extract the maximum amount of information of the system in the fewest number of runs. Design of experiments study the influence of several factors in order to optimize processes and learn the relationships between the factors over a wide range of values and how they affect the response of the process environment. The most important stage in DoE process is the screening experiments. They consist in the realization of only a few experiments to find out the most relevant information about the process. There are several different types of screening designs but the common one is the fractional factorials design. The screening approach based on two-level designs is the most basic approach but it is sufficient to estimate linear and interaction models. Then, the full factorial design executes a set of experiments where every level of the factor is observed at both levels of all the other factors (see Fig. C.6). For example, with n factors and L levels it is executed L^n experiments. As an alternative, a fractional factorial design approach runs a subset of the full experiments without the loss of too much information. With fractional factorial design, 3-way and higher interactions are neglected. One popular method to produce fractional factorial design in industrial experiments is given by orthogonal arrays often referred to as *Taguchi Methods* [24].

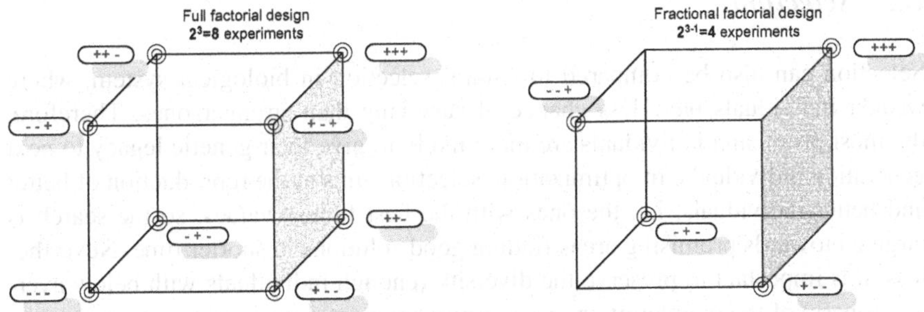

Fig. C.6 Example of full and fractional design for three levels experiments

Latin hypercube sampling (LHS) [26] is a form of stratified sampling that generates a more even distribution of parameter values in the multidimensional space than typically occurs with pure *Monte Carlo* (MC) sampling. Variables are sampled using a square grid symmetrically arranged allowing only one sample in each row and each column. In two dimension variables this structure is called a Latin square, as illustrated in Fig. C.7. When extrapolated for multi-dimension spaces, it is called a Latin hypercube where only one sample is admitted in each axis, aligned with the hyper plane containing it.

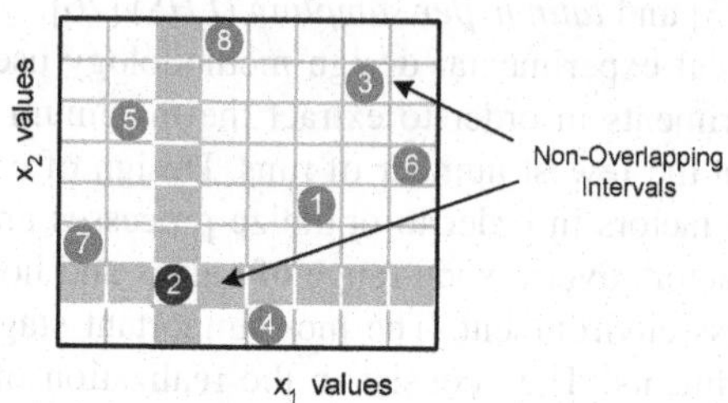

Fig. C.7 A Latin Hypercube Sample with two variables and eight even intervals

When sampling the space of *N* variables (*X1*,..,*Xn*), the range of each variable is divided into *K* equally intervals. Then, one value from each interval is selected at random with respect to the probability density in the interval. The n values achieved by *X1* are paired randomly with the n values of *X2*, *X3* and so on, until *n* *k*-tuplets are formed.

One of the advantages of this sampling scheme is that the number of sample points is independent of dimension of problem. It is not necessary to take more sampling points when the dimension of the problem increases. Moreover, the particular grid structure allows the remembrance of the last random samples.

C.5 Selection

Selection can also be compared to natural selection in biological system, where weaker individuals have less chance of surviving than stronger ones. Therefore, the most promising individuals are more likely to give their genetic legacy to next generation individuals. In optimization, selection aims at the reproduction of better and better individuals, i.e. the ones with the best fitness values, so the search is targeted towards promising areas finding good solutions in shorter time. Nevertheless, it is important to preserve the diversity (enough individuals with below average fitness) of the population in order to prevent premature convergence and at the same time provide enough selective pressure (rate of individuals with above average fitness) to allow the population convergence to the global optimal solution.

Several selection algorithms were developed to provide the harmony between these two antagonistic activities, selective pressure and diversity. Table C.5 describes the most common selection methods.

Table C.5 Selection operators

Methods	Description	Advantages/Disadvantages
Roulette wheel [21]	Resembles the functioning of a real roulette wheel, where fitness values of individuals correspond to the widths of slots on the wheel machine. Higher fitness values represented in wider slots are more likely to be chosen to next generation when a random selection is initialized	As soon as the population converges upon solution, selective pressure decreases severely affecting the search of better solutions.
Stochastic universal sampling (SUS) [21]	A small variation of roullete wheel method. SUS provides a fitness-proportionate selection with minimal use of a stochastic process. Instead of spinning a roulette wheel one time for each n number of offsprings, the roulette wheel is spinned with n equallyspaced pointers just once.	SUS is optimally more efficient than roulette wheel.
Tournament Selection [27]-[28]	One parent is selected randomly, comparing the fitness of n individuals in the actual population and selecting the fittest. The second parent is selected by repeating the same process. The binary tournament selects the parents using two (n is equal to two) competitors.	This type of selection allows the control of the selection pressure rate. And is easy to implement. Is convenient to compare the performance of individuals.

The selection criteria are very general and different methodologies of selection schemes can be applied:

1. In a *generational* selection, the entire population can be replaced by the new offspring. This method does not guarantee that that best individual will be part of the next generation.
2. The *elitism* selection, on the contrary, implements a mechanism that copies the best individual to the next generation unconditionally. Here, only a subset of the original population is replaced, in this case, the algorithm is called a steady-state EA.
3. The last general scheme is the *sharing* or *crowding* selection that was proposed for the optimization of multimodal functions. Here, the objective is to maintain a population distributed over all or many of the optima regions. This behavior is normally achieved by reducing the fitness value of an individual in dense regions (crowd) according to some "similarity" metric. This encourages the search in unexplored regions and causes the appearance of subpopulations. A problem with sharing methods is the introduction of two new parameters: a new sharing criterion and the need to define the "similarity" metric.

C.6 Crossover Operator

The recombination or crossover operator is the main search operator in the GAs. The aim of the crossover operation is to produce offspring that have large fitness values, satisfying the problem's constraints. The most common techniques to

Table C.6 Crossover operator overview

Crossover Methods	Description	Advantages/Disadvantages
N-point [29]	Defines N random crossing points to exchange gene information, with N=(1,..n)	Classical approach. Can be used with categorical data.
Uniform [27]	Each gene in the offspring chromosome decides (with probability p) which of the two parents will contribute with its genetic information to form the mutated gene in that position.	It is possible to combine different characteristics independently of the relative position in the chromosome.
Arithmetic [23]	This operator is a linear combination of two vectors (chromosomes): let x1 and x2 be the parents selected to breed, then the resultant offsprings will be given by $x_1^{'} = \lambda.x_1 + (1-\lambda).x_2$ and $x_2^{'} = (1-a).x_1 + a.x_2$ where λ is a random number between [0,1].	This operator is particularly suitable for numeric problems with constraints where the feasible region is convex.
Heuristic [23]	Produces a single offspring through linear extrapolation between two individuals. Let x and y be the two parents selected to breed then final offspring will be given by the expression $z = \lambda.(y-x) + y$. If the generated solution is not feasible a new random number is created.	Exploit the "quasi-gradient" of the evaluation function as a means of directing the search process.
Mean Centric [33]	The mean-centric recombination groups a class of operators that produce offspring near the centroid of the involved parents. Examples of these techniques include, unimodal normal distribution crossover (UNDX), simplex crossover (SPX) and blend crossover (BLX).	Can be useful for exploration purposes.
Parent Centric [33]	In parent-centric recombination, offspring are created in the vicinity of the parents. It is given to each parent an equal probability of creating offspring in its neighbourhood such as parent-centric recombination operator (PCX)	Can be useful for exploitation purposes.

implement the crossover operations are those derived from classical evolution theory like the one-point crossover, N-point crossover [29], the uniform crossover [27], the arithmetic and the heuristic crossover [23], etc. Table C.6 reviews a few generic (problem independent) crossover operators found in literature.

The crossover operator generates new individuals (offspring) through the recombination of two or more parents. Crossover can be compared to sexual reproduction in natural organisms as it permits the swapping of information between individuals.

A different approach is given by the EDAs (section 3.2.3) algorithms which employ probabilistic models of the search distribution that model crossover operators. These methods introduce the idea of correlated exploration to the field of recombination algorithms. However, EDAs are not efficient to the continuous optimization [30]-[32].

C.7 Mutation Operator

The mutation operator is the primary variation/search operator in ESs while in GAs it is often considered a useful complement of crossover, usually performed with a low probability [21]. The main role of mutation in GA is to assure the diversity of genetic information in the population in order to prevent the premature convergence of GA to sub-optimal solutions. In practice, mutation changes the value of individual genes at random with a certain probability and assures that all the points in the search space are likely to be examined. The probability of occurring a mutation in a gene is called the mutation rate.

The GAs typically employ only one mutation rate pm for the population. Generally, the mutation rate value is fixed, not allowing any change or self-adaptation during evolution. Table C.7 describes some of commonly used mutation techniques.

The mutation operator plays an important role in applications of adaptive parameter control or self-adaptation principles in evolutionary algorithms. In adaptive parameter control, the parameter settings (involving mutation and sometime crossover) attain different values according to a deterministic or probabilistic schedule defined by the user, for example, varying the mutation rate over the number of generations of the algorithm.

Table C.7 Mutation operator techniques

Mutation Methods	Description	Advantages/Disadvantages
Standard [21]	This type of operator just complements the binary value of the gene selected for mutation.	Limited to binary operators
Uniform [21]	Choose the component to mutate, and then change this component value by a random number sampling inside the limits of parameter x=[lb,ub] where lb and ub means the lower and upper bound.	The admissible values applied to real valued genes can take any statistical pattern
Gaussian [21]	Now the (real) component value of individual x_k is changed to $x_k^{'} = x_k + N(0,\sigma)$ by a random value obtained from a gaussian distribution $N(0,\sigma)$ of mean zero and standard deviation σ.	The parameter σ is user defined and should be carefully chosen. This approach can be used with adaptively mechanisms.

The self-adaptation concept, which evolved from evolution strategies and evolutionary programming techniques, changes the value of mutation online, during the search, by applying the search operator(s) mutation (and recombination, in case of evolution strategies) to the optimization parameters. This method incorporates the control parameters into the chromosome.

C.8 Performance Criteria

The two most used metrics to measure the performance of an algorithm are the effectiveness and efficiency [35]. *Effectiveness* measures the capacity of the algorithm to accomplish the objectives. This value can be calculated by empirical observation in bunch of tests functions measuring the number of times the optimum has been reached by a certain algorithm. In case of algorithms with stochastic ground, the performance criterion is measured as the average of repetition trials. *Efficiency* is the effort needed by the algorithm to reach the optimum. In evolutionary algorithms, it is the number of function evaluations or number of generations consumed to reach the target. Other aspects that could be relevant in certain cases can include a metric to trace down the performance of the algorithm in terms of the number of feasible solutions found or even the convergence rate as well.

The classic way used to study the performance of an algorithm is through a performance graph showing the trade-off between the two main criteria or making use of tables comparing the performance of one or several algorithms against several parameter settings and running over some test functions during a predefined number of function evaluations.

Appendix D. Support Vector Machine Overview

D.1 The SVM Model Formulation

The classical two class classification case defined by a set of training data of the form $S=\{(x_i; y_i);\ldots;(x_n; y_n)\}$, where the input $x_i \in X \subset R^d$ is a d-dimensional feature vector and the output $y_i \in \{+1,-1\}$ is the class label of x_i.

In the first implementation step, SVM applies the kernel "trick", which provides a nonlinear mapping of the vectors x_i into a higher dimensional feature space. Mathematically, it can be described as a nonlinear mapping ϕ, $\phi : \Re^n \rightarrow \mathrm{H}$, where H is a high dimension dot metrics space entitled *Hilbert space* or *feature space*, and $\phi(x)$ the feature mapping. For nonlinear problems, the two classes are more easily separated in H than in R^d. ϕ must be chosen so that the kernel operator $K(x, x') = <\phi(x), \phi(x')>H$ is positive definite. This allows us to compute inner products in H without explicitly evaluating ϕ [35].

In the second step, a decision boundary hyperplane is created based on the maximal-margin principle as illustrated in Fig. D.1.

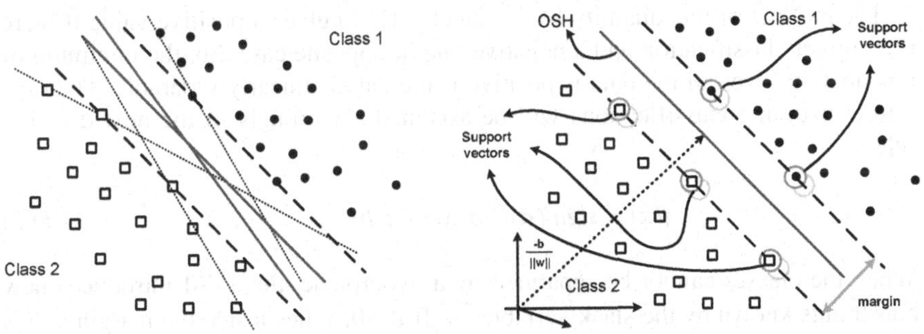

Fig. D.1 Illustration of the main SVM concepts

The decision boundary points overlapping the margins are called support vectors. Between them, an infinite number of separating hyperplanes are admissible (Fig. D.1, left) including the optimal separating hyperplane (OSH) [36] of two separable classes (Fig. D.1, right).

The distance from the origin to the optimal separating hyperplane is given by $(-b/||w||)$, where $\boldsymbol{w}$ is the normal vector of the hyperplane whose norm is held constant and $\boldsymbol{b}$ a real number offset parameter often called the ***bias***.

The margin M is given by the following quantity:

$$M = min_i\, y_i\{<w, \phi(xi)> + b\} \quad (D.1)$$

where < , > denotes an inner product, the hyperplane is defined by $\boldsymbol{w}$ and $\boldsymbol{b}$ and the expression given by $(<w, \phi(xi)> + b)$ corresponds to distance between the point x_i and the decision boundary (see Fig. D.2).

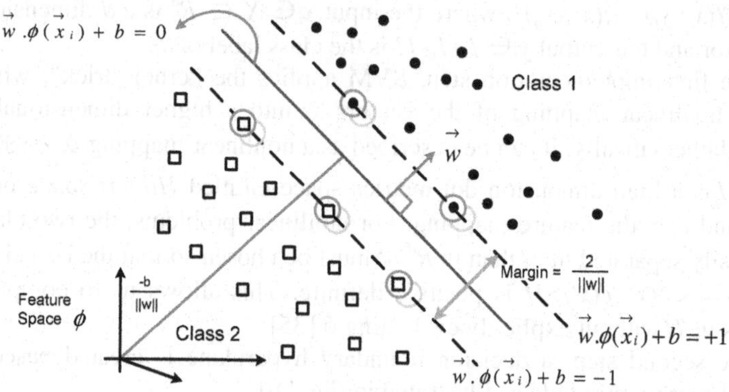

Fig. D.2 Margin and hyperplane - Mathematical expressions

The product of this quantity by the label y_i (D.1) gives a positive value if there is a correct classification and a negative one in opposite case. So, the minimum of this quantity over all the data is positive if the data is linearly separable. Then the future incoming classifications will be assigned accordingly to the next decision rule:

$$f(x) = sign\,(<w, \phi(xi)> + b) \quad (D.2)$$

When the classes cannot be separated by a hyperplane, the SVM introduces new constraints known by the slack variables ε_i. If $\varepsilon_i > 0$, x_i lies inside the margin and is called a margin error. The distance between the hyperplane and misclassification is given by $(-\xi i/\|w\|)$ [36].

Finally, the SVM can be formulated as the following quadratic program:

$$\min_{w,b,\varepsilon} \tfrac{1}{2}\|w\|^2 + C\sum_{i=1}^{n} \varepsilon_i \quad (D.3)$$

Subject to,

$$y_i(k(w, x_i) + b) \geq 1 - \varepsilon_i; \quad \text{for } i=1,\ldots,n$$
$$\varepsilon_i \geq 0$$

and, $C \geq 0$ is a parameter that controls the tradeoff between minimizing the margin errors and maximizing the margin.

For computational reasons, it is often easier to solve the equivalent dual problem using the Lagrangian formulation (αi is the Lagrange multipliers):

$$\min_{\alpha} \tfrac{1}{2} \sum_{i,j=1}^{n} \alpha_i \alpha_j y_i y_j k(x_i, x_j) - \sum_{i=1}^{n} \alpha_i$$

$$\text{Subject to,} \tag{D.4}$$

$$0 \leq \alpha_i \leq C \qquad \text{for i=1,...,n}$$

$$\sum_{i=1}^{n} \alpha_i y_i = 0$$

The primal and the dual are related through $w = \sum_{i=1}^{n} \alpha_i y_i \phi(x_i)$. Usually $\alpha i = 0$ for most of the x_i. The points x_i that have non-zero Lagrange multipliers α_i are termed the Support Vectors (SV). If the data are linearly separable, all the SVs will lie on the margin and hence the number of SVs can be very small (Fig. D.1).

The kernel function performs the non-linear mapping into the feature space. The choice of kernel to fit non-linear data into a linear feature space depends on the structure of the data [36]. Some of the most popular kernels which are used in most SVM packages are presented in Table D.1.

Table D.1 Typical SVM kernels.

1. Linear kernel:	$K(x, y) = \langle x^T y \rangle$
2. The Radial Basic Function kernel where, the kernel width is user-defined.	$K(x, y) = \exp(-\|x - y\|^2 / 2\sigma^2)$
The polynomial kernel where, the degree of the polynomial, d, is also user-defined.	$K(x, y) = (\langle x^T y \rangle + 1)^d$
4. Sigmoid with parameter κ and θ	$K(x, y) = \tanh(\kappa x^T y + \theta)$

D.2 Data Setup

This is the data preparation step before building the model. This step involves the identification and normalization of data samples.

D.2.1 Data Collection

The data samples needed to build the model are collected in a database. This is considered the most time consuming task of the overall process, since data sampling usually evolves the collection of large number of expensive process samples. The use of database management systems (DBMS) may help the exploration and extraction of information in order to understand this data process. However, in other situations (depending on the amount and the complexity of the data) a flat file or even a spreadsheet may be adequate [37].

D.2.2 Pre-processing of the Training Data

The SVM algorithm operates on numeric attributes and it is applied in a variety of domains. The relationship between the object under study and its attributes can take multiple representations, be stored in several data structures even using different data types formats. Therefore, a common normalization is desired to achieve the data representation required by the SVM. For example, in applications where categorical data (non real data) is available, a transformation of the categorical in binary format is required. These early stage normalizations are application specific so a couple of these were implemented within GENOM, as described. First of all, the source of I/O SVM data is done in text data files, allowing the efficient sparse data representation and storing a single object-attribute pair in each line. Second, a normalization procedure (placed on similar scale) is submitted to each individual data attributes. This step prevents attributes with a large original scale from biasing the solution preventing eventual computational overflows and underflows. This is achieved by scaling the training data to a predefined range normally between [0, -1] or [-1, -1]. The scaling routine reads through the training data file to determine the maximum and minimum for each component of the training vector. Then, values for the same component of all examples are linear scaled according to the following equation:

$$ScaleValue = Lower + (Upper - Lower) * \frac{OriginalValue - MinimumValue}{MaximunValue - MinimumValue} \quad (D.5)$$

The maximum and minimum values of each component are saved in a SVM description file to avoid referring the training data again when scaling the testing data. The description file makes the management and update of the extreme values for scaling in streaming events easy. Moreover, during the scaling phase the existence of unwanted outliers and long tailed distributions can produce bad resolution scale intervals and should also be prevented.

D.2.3 Unbalanced Data Sets

Real world applications are often characterized by highly unbalanced data distributions. The ratio of positive to negative examples is small, meaning that one class is under-represented compared to the other. Frequently, the class with more interest to the user (the positive training samples) is represented by the minority class. This scenario improperly biases the classifier and can significantly reduce the accuracy of a classifier. The SVM models trained in such conditions will tend to predict the majority class [38]-[39].

The main activities for handling unbalanced data problems are focusing in two main methods that alter the class distribution of data sets: the under-sampling, which shrink the size of samples in the majority class, and over-sampling, which increase the number of samples in the minority class. The basic methods employ a random sampling for reducing the majority samples and the replication of samples in the case of over-sampling. The overall effect is to diminish the high class asymmetries in the training set.

Whether it is used individually or simultaneously, both methods have some drawbacks. For example, the under-sampling method can throws away potentially useful samples near the decision boundary. Those samples could be potentials support vectors responsible for the accuracy of the model. On the other hand, replicating the majority class examples increases the size of the training set increasing the cost to build the model (method used alone) and may also lead to overfitting models. Recent studies have emerged which tries to optimize the efficiency and accuracy of the model [40]-[41] and [42].

Other methods exist to improve the accuracy and performance of the learning model without changing the class distributions for unbalanced problems. They are based in the principle that the error introduced by a wrong estimation or classification have different significance for different classes. The cost introduced by a wrong classification of the interesting class sample (a minority) has greater impact than the cost introduced by a wrong classification of the majority class sample, that's why they are called cost-sensitive methods. Thus, the cost-sensitive learning methods, belongs to the class of classifiers that minimize cost as well as the traditional error rate whose impact of the costs weights can be parameterized by the user in a class basis. Assign a higher weight factor or cost C to the minority samples in detriment of less cost values for majority classes, assures a biasing model that gives more "representation" to small classes. Now, the SVM formulation presented in expression D.3 suffer a slight modification and became:

$$\min_{w,b,\varepsilon} \tfrac{1}{2}\|w\|^2 + C_1 \sum_{i=1}^{\#negative} \varepsilon_i + C_2 \sum_{i=1}^{\#positive} \varepsilon_i \qquad (\text{D.6})$$

One example to accomplish this is defined in [43], and assigns the costs values to C1 and C2 to each class as below:

$$C_1 = \tfrac{m2}{M} C \text{ and } C_2 = \tfrac{m1}{M} C \qquad (\text{D.7})$$

Where, *m1* and *m2* denote the size of class 1 and class 2 data sets, respectively, with *m1*>>*m2* and *M*=*m1*+*m2*. Given a higher cost term to the minority class *C2* will produce a model that can better predict new data.

D.3 SVM Model Building

A well defined training and validation procedure is required in order to insure accurate and robust predictions. However, the limited availability of data resources in some cases or the high cost of data collecting process may impose difficult challenges to obtain the true accuracy. An accurate estimation of the true accuracy should be put into practice for small data set cases. The quality of the estimation also depends on the methods to measure the performance of the clas-sifier. The most common used methods described next are the training and testing, the bootstrap and cross-validation approach.

D.3.1 Training and Testing by Simple Validation Approach

In the *simple validation* method, the search space *S* is randomly partitioned into two subsets *S1* and *S2* with asymmetrical loads, generally *S1* with 2/3 of the total data and *S2* with the rest 1/3. A model is built, using as training set *S1* and the accuracy tested with the *S2* subset. This process is repeated *N* times with different random partitions, then the true accuracy is obtained averaging the results of each iteration. It provides a way of evaluating the performance of a model trained with the given training parameters. This method behaves well for large data sets. Small data sets usually lead to inaccurate estimation with large bias because a significant portion of data (*S2*) was spent to represent the test data. Another weakness of this technique is that it violates the requirements for independence of test sets because the partition of the test sets is not disjointed [44].

D.3.2 Bootstrap Method

The bootstrap method is a technique for estimating the error of a model. The bootstrap method generates *N* subsets (S_1, S_2, …, S_N) from the original set *S* using a sampling technique usually based on a random with replacement strategy. The basic process chooses randomly one element of the entire set *S*, adds it to S_i and puts a copy back into *S* (replacement). This process is repeated *T* times equal to size of elements of *S*. As a result, the total number of elements of subsets S_i is equal to *T*. There is also a probability that several elements of *S* can be copied several times to S_i while some may have none. In this case the classifier is built using S_i as the training set while the test set is formed by all elements of *S* not included in S_i. The final accuracy model is obtained by averaging the accuracy in each subset.

D.3.3 Cross-Validation Method

Considered as one of the most reliable but also most expensive in terms of computational cost, *N-fold cross-validation* randomly divides the training data into *N* sets [44]-[46]. Then it builds *N* models, each time leaving one of the set out as the testing set. Again, the average accuracy rate is calculated for each fold. A particular feature of this method is that all the test set is disjoint and thus each training set is tested only once. Briefly, the *N-fold cross validation* algorithm is described in three main steps (Fig. D.3):

```
1. Divide the training set (of size m) into n disjoint sets
   S1;S2;...;Sn of equal size n/m.
2. For each Si:
      - Train a classifier on S\Si
      - Test it on Si —> error(i)
3. Output the average error
```

Fig. D.3 N-fold cross validation algorithm

This gives an estimate of the generalization error of the classifier when trained on *n-n/m* data. Usually *n* is equal to 10. When applied to model selection parameter, a *leave-10-out cross-validation* is often used applying the following sequence of actions:

- For each set of values of the parameters, leave-10-out cross-validation on the training set is performed to estimate the models accuracy.
- Select the set of values from each parameter that produced the model, which gave the smallest prediction error (optimal parameter settings).
- Once a good set of parameter values is found, train the model with the optimal parameter settings for whole training set and test it with a test set (test is not used for training).

D.4 SVM Model Evaluation

After the building process, the performance of the predictive model is normally measured with a set of evaluations metrics. Based on these results, the model may need to be rebuilt again using a different technique or the same technique with a different set of samples in order to increase the model accuracy.

D.4.1 Kernel Evaluation Metrics

The performance of a classifier is achieved mainly by the measure of the true accuracy obtained with the original data set *S*. There are several performance measures used to estimate the accuracy or quality of the model.

1. The *Root-Mean-Square-Error* (RMSE) is often used as a performance criterion in cross-validation and also for predicting the test set. The RMSE value is defined by:

$$RMSE = \sqrt{\frac{\sum_{i=1}^{n}(\hat{y}_i - y_i)^2}{n}} \qquad (D.8)$$

   where, $\hat{y}i$ and yi are the predicted and real values of sample i of the n samples respectively.
2. The Receiver Operating Characteristic (ROC) curve, illustrated in Fig. D.4, is a graphical technique used to visualize the relation between true and false positives. The axes of a ROC curve are the number of true positives divided by the total positives in the test set and the false positives divided by the total number of negatives. The Area Under the Curve (AUC) gives a scalar measurement for the performance. An AUC value of '1' represents a perfect test; an area of '0.5' represents a worthless test. Fig. D.4 illustrates the ROC concept.
3. The *precision* and *recall* metrics are two useful measures for evaluating the quality of results in statistical classification problems. The Precision is the percentage of the outcome of a statistical task matching the desire results. In the classification task, precision is the number of true positives (i.e. the number of items correctly labeled as belonging to the class) divided by the total number of

elements labeled as belonging to the class (i.e. the sum of true positives and false positives, which are items incorrectly labeled as belonging to the class). In the same context, Recall is defined as the number of true positives divided by the total number of elements that actually belong to the class (i.e. the sum of true positives and false negatives).

$$\text{Precision} = \frac{TP}{TP+FP} \quad ; \quad \text{Re}\,call = \frac{TP}{TP+FN} \tag{D.9}$$

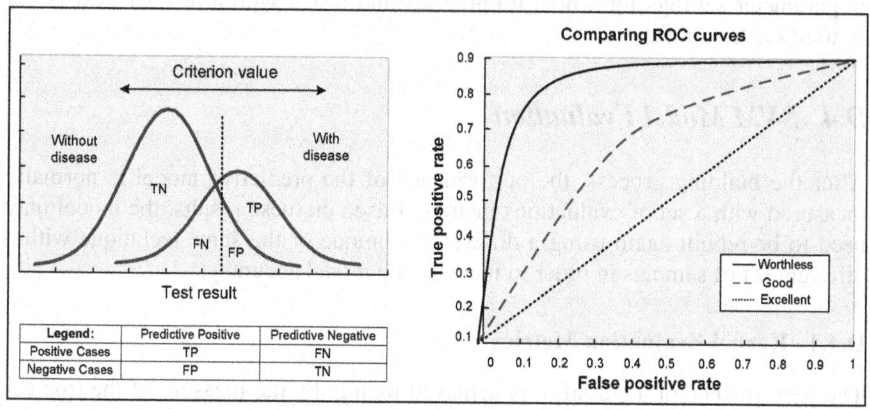

Fig. D.4 Receiver operating characteristic (ROC) curve

D.4.2 Model Selection Parameters

Model selection also known as parameter tuning is one of the most critical steps in the process of building SVM models. It is well known that SVM generalization performance depends on a good set of parameters C, ε and the kernel setting (σ in case of RBF or d in case of polynomial kernels), which must be defined by the user. An inappropriate choice of these parameters may lead to *underfitting* (e.g., for classification, the model always predicts the dominant class), *overfitting* (i.e., the model memorizes the training data) or slow and inefficient models. Fig. D.5 illustrates the influence of these parameters in SVM modeling.

There are several methods found in the literature for tuning the SVM parameters [36], [44] and [47]-[52]. The aim of these techniques is to find the optimal parameters that minimize the prediction error of the SVM model. A common practice is to use a grid search approach to find the optimum values. In this case, a grid is span all over the search space of the parameters models. The grid resolution has great impact on the quality of the model and the time consumed in this process. To alleviate some of these drawbacks, sometimes the tuning process implements a coarse grid search followed by a fine grid approach in the regions of the most promising regions identified in the first step. Several other methods were proposed to speed up the grid search approach, including the use of optimization methods. Some approaches employ the design of experiments techniques [48], pattern search algorithms and stochastic algorithms [51]-[52].

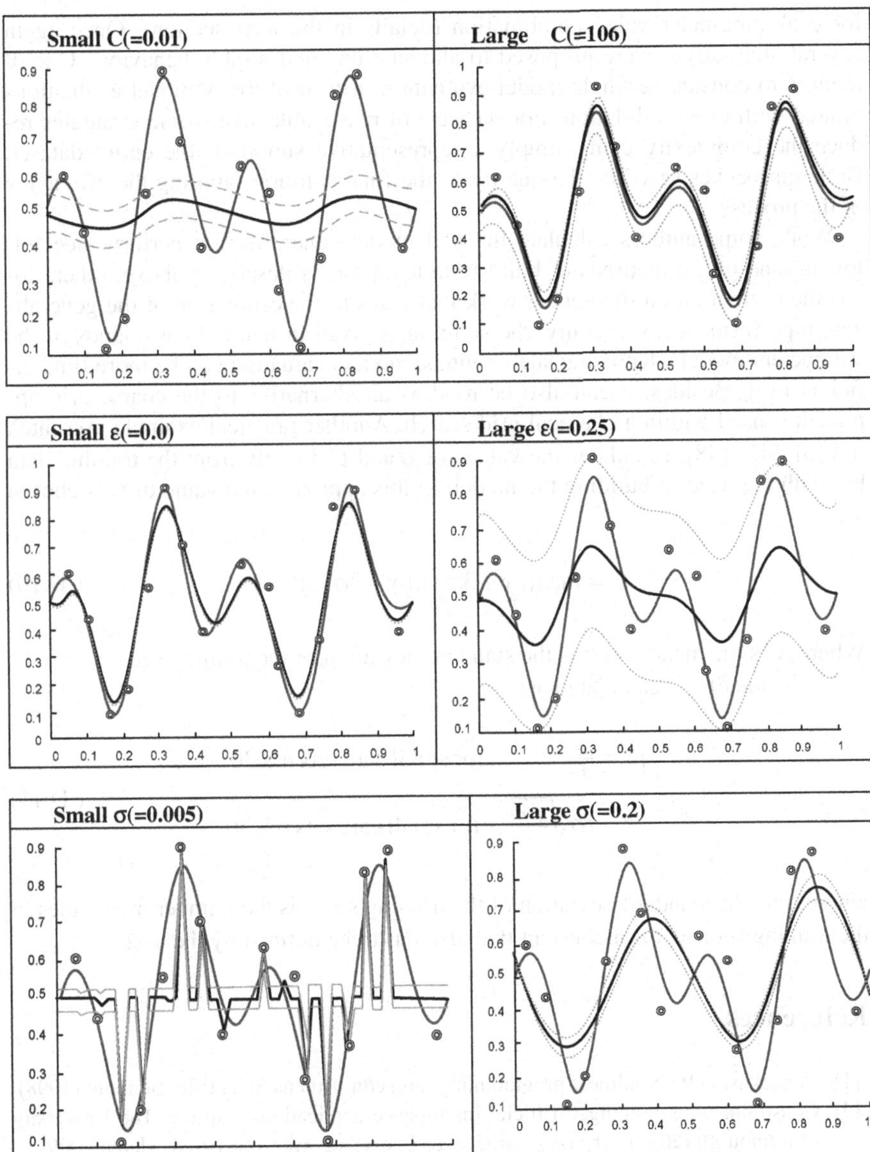

Fig. D.5 Influence of the hyperparameters on SV regression [46]

Whichever the process used, a measure of the model quality should be adopted to measure the overall generalization performance. *N-fold cross-validation* is considered one of the most confident methods for this purpose. However, it is also a time consuming process because it is coupled with the generation of *N+1* model for each parameter value combination (details in the next section). Once again several alternatives were proposed to alleviate this undesirable behavior. One of them is to consider a single model evaluation instead of the *N* model evaluations related with cross-validation. For datasets of reasonable size, some strategies reduce the complexity using simply a representative subset of the entire dataset. Both approaches sacrifice the quality of the final solution favoring the efficiency of the process.

While some authors calculate the real model generalization performance following one of the methods of build train-test process described above, others focus their attention on theoretical work that leads to the estimation of the generalization performance. Generally, the solution is given in intervals or bounds of the parameters models. Knowing such bounds, model evaluation can be more time efficient [35]. Besides, it can also be used as an alternative to the coarse grid approach if used within a classical grid search. Another practical example presented in [36], [47]-[48], calculates the values of σ and C directly from the training data but with the need of building the model. In this approach, the value of C is chosen as:

$$C = \max(|\underline{y} + 3\sigma_y|, |\underline{y} - 3\sigma_y|) \tag{D.10}$$

Where, y is the mean and σ is the standard deviation of the training set.

The value of ε is calculated as:

$$\begin{cases} \varepsilon = \frac{\sigma}{\sqrt{n}} & \text{, for small data sets n} \leq 30 \\ \varepsilon = \tau\sigma\sqrt{\frac{\lg(n)}{n}}, & \text{for small data sets n} \geq 30 \end{cases} \tag{D.11}$$

where, σ is the standard deviation of the training set, n is the number of samples in the training set and τ is a constant that also has to be defined by the user.

References

[1] Bertsekas, D.P.: Nonlinear programming, 2nd edn. Athena Scientific, Belmont (1998)

[2] Constraint programming, Artificial intelligence applications institute. The University of Edinburgh (2007), http://www.aiai.ed.ac.uk/ (accessed, March 2009)

[3] Deb, K.: An efficient constraint handling method for genetic algorithms. Computer Methods in Applied Mechanics and Engineering, vol. 186, pp. 311–338. Elsevier, Amsterdam (2000)

[4] Coello, C.A.C.: Theoretical and numerical constraint handling techniques used with evolutionary algorithms: A survey of the state of the art. Computer Methods in Applied Mechanics and Engineering, vol. 191, pp. 1245–1287. Elsevier, Amsterdam (2002)

[5] Mezura-Montes, E., Velázquez-Reyes, J., Coello, C.A.C.: Promising infeasibility and multiple offspring incorporated to differential evolution for constrained optimization. In: Proc. GECCO, pp. 225–232 (2005)
[6] Mahfoud, S.W., Goldberg, D.E.: Parallel recombinative simulated annealing: A genetic algorithm. Parallel Computing 21, 1–28 (1995)
[7] Goldberg, D.E.: Genetic algorithms in search, optimization and machine learning. Addison-Wesley, Reading (1989)
[8] Nelder, J.A., Mead, R.: A Simplex Method for Function Minimization. Computer Journal 7, 308–313 (1965)
[9] Metropolis, N., Rosenbluth, A.W., Rosenbluth, M.N., Teller, A.H., Teller, E.: Equation of state calculations by fast computing machines. Journal of Chemical Physics 21(6), 1087–1092 (1953)
[10] Kirkpatrick, S., Gerlatt, C.D., Vecchi, M.P.: Optimization by simulated annealing. Science (1983), doi:10.1126/science.220.4598.671
[11] Schaffer, J.D.: Some experiments in machine learning using vector evaluated genetic algorithms. Ph.D. dissertation, Vanderbilt University, Nashville, TN (1984)
[12] Fonseca, C.M., Fleming, P.J.: Genetic algorithms for multi-objective optimization: Formulation. Discussion and Generalization. In: Proc. 5th International Conference on Genetic Algorithms, pp. 141–153 (1993)
[13] Fonseca, C.M., Fleming, P.J.: Multi-objective optimization and multiple constraints handling with evolutionary algorithms–Part II: Application example. IEEE Trans. Systems, Man, and Cybernetics: Part A: Systems and Humans, 38–47 (1998)
[14] Knowles, J., Corne, D.: The pareto archived evolution strategy: A new baseline algorithm for multi-objective optimization. In: Proc. Congress on Evolutionary Computation, pp. 98–105. IEEE Service Center, Piscataway (1999)
[15] Deb, K., Pratap, A., Agrawal, S., Meyarivan, T.: A fast and elitist multi-objective genetic algorithm: NSGA-II. IEEE Trans. Evolutionary Computation 6, 182–197 (2002)
[16] Horn, J., Nafploitis, N., Goldberg, D.E.: A niched pareto genetic algorithm for multi-objective optimization. In: Proc. 1st IEEE Conference on Evolutionary Computation, pp. 82–87 (1994)
[17] Zitzler, E.: Evolutionary algorithms for multi-objective optimization: Methods and applications. Ph.D. Thesis, Swiss Federal Institute of Technology (ETH), Zurich (1999)
[18] Fonseca, C.M., Fleming, P.J.: An overview of evolutionary algorithms for multi-objective optimization. In: Proc. Congress on Evolutionary Computation, vol. 3(1), pp. 1–16 (1998)
[19] Holland, J.H.: Adaptation in Natural and Artificial Systems. The University of Michigan Press, Ann Arbor (1975)
[20] Eiben, A.E., Hinterding, R., Michalewicz, Z.: Parameter control in evolutionary algorithms. IEEE Trans. Evolutionary Computation 3(2), 124–141 (1999)
[21] Bäck, T., Fogel, D., Michalewicz, Z.: Handbook of evolutionary computation. Oxford Univ. Press, Oxford (1997)
[22] Czarn, A., MacNish, C., Vijayan, K., Turlach, B., Gupta, R.: Statistical exploratory analysis of genetic algorithms. IEEE Trans. Evolutionary Computation 8(4), 405–421 (2004)
[23] Michalewicz, Z.: Genetic algorithms + data structure = evolution programs, 3rd edn. Springer, Berlin (1996)

[24] Antony, J., Somasundarum, V., Fergusson, C.: Applications of taguchi approach to statistical design of experiments in Czech Republican industries. International Journal of Productivity and Performance Management 53(5), 447–457 (2004)

[25] Trygg, J., Wold, S.: Introduction to statistical experimental design. Editorial (2002), http://www.acc.umu.se/~tnkjtg/Chemometrics/Editorial (accessed, March 2009)

[26] McKay, M.D., Conover, W.J., Beckman, R.J.: A comparison of three methods for selecting values of input variables in the analysis of output from a computer code. Technometrics 21, 239–245 (1979)

[27] Miller, B.L., Goldberg, D.E.: Genetic algorithms, tournament selection, and the effects of noise. Illinois Genetic Algorithms Laboratory, Tech. Rep. TR No: 95006 (1995)

[28] Legg, S., Hutter, M., Kumar, A.: Tournament versus fitness uniform selection. In: Proc. Congress on Evolutionary Computation, vol. 2, pp. 2144–2151 (2004)

[29] Haupt, R.L., Haupt, S.E.: Practical genetic algorithms. Wiley, New York (1998)

[30] Ocenasek, J.: Parallel estimation of distribution algorithms. Ph.D. dissertation, Faculty of Information Technology, Brno University of Technology (2002)

[31] Larrañaga, P., Lozano, J.A.: Optimization by learning and simulation of probabilistic graphical models. In: Parallel Problem Solving from Nature, PPSN VII (2002), http://www.sc.ehu.es/ccwbayes/ (accessed, March 2009)

[32] Larrañaga, P., Lozano, J.A.: Estimation of distribution algorithms: A new tool for evolutionary computation. Kluwer Academic Publishers, Norwell (2001)

[33] Raghuwanshi, M., Kakde, O.: Survey on multiobjective evolutionary and real coded genetic algorithms. In: Proc. 8th Asia Pacific Symposium on Intelligent and Evolutionary Systems, pp. 6–10 (2004)

[34] Baritompa, W.P., Hendrix, E.M.T.: On the investigation of stochastic global optimization algorithms. Journal of Global Optimization 31, 567–578 (2005)

[35] Milenova, B.L., Yarmus, J.S., Campos, M.M.: SVM in oracle database 10g: Removing the barriers to widespread adoption of support vector machines. In: Proc. 31st International Conference on Very Large Data Bases, pp. 1152–1163 (2005)

[36] Üstün, B.: A comparison of support vector machines and partial least squares regression on spectral data. Master thesis, University of Nijmegen, The Netherlands (2003), http://www.cac.science.ru.nl/people/ustun (accessed, March 2009)

[37] Edelstein, H.A.: Introduction to data mining and knowledge discovery, 3rd edn. Two Crows Corporation (2003), http://www.twocrows.com/intro-dm.pdf (accessed, March 2009)

[38] Liu, Y., An, A., Huang, X.: Boosting prediction accuracy on imbalanced datasets with SVM ensembles. In: Ng, W.-K., Kitsuregawa, M., Li, J., Chang, K. (eds.) PAKDD 2006. LNCS (LNAI), vol. 3918, pp. 107–118. Springer, Heidelberg (2006), doi:10.1007/11731139

[39] Liu, A.: The effect of oversampling and undersampling on classifying imbalanced text datasets. Master thesis, University of Texas, USA (2004)

[40] Romano, R.A., Aragon, C.R., Ding, C.: Supernova recognition using support vector machines. In: Proc. 5th International Conference on Machine Learning and Applications, pp. 77–82 (2006)

[41] Brank, J., Grobelnik, M., Milic-Frayling, N., Mladenic, D.: Training text classifiers with SVM on very few positive examples. Tech. Rep. MSR-TR-2003-34 (2003)

[42] Akbani, R., Kwek, S., Japkowicz, N.: Applying support vector machines to imbalanced datasets. In: Proc. European Conference on Machine Learning, pp. 39–50 (2004)

[43] Shina, H., Cho, S.: Response modeling with support vector machines. Elsevier - Expert Systems with Applications 30(4), 746–760 (2006)

[44] Zeng, X., Martinez, T.R.: Distribution-balanced stratified cross-validation for accuracy estimation. J. Experimental & Theoretical Artificial Intelligence 12(1), 1–12 (2000)

[45] Duan, K., Keerthi, S., Poo, A.: Evaluation of simple performance measures for tuning SVM hyperparameters. Neurocomputing 51, 41–59 (2003)

[46] Ito, K., Nakano, R.: Optimizing support vector regression hyperparameters based on cross-validation. In: Proc. International Joint Conference on Neural Networks, vol. 3, pp. 2077–2082 (2003)

[47] Kiely, T., Gielen, G.: Performance modeling of analog integrated circuits using least-squares support vector machines. In: Proc. Design, Automation and Test in Europe Conference and Exhibition, vol. 1, pp. 448–453 (2004)

[48] Cherkassky, V.L., Ma, Y.: Practical selection of SVM parameters and noise estimation for svm regression. Neural Networks 17(1), 113–126 (2004)

[49] Frohlich, H., Zell, A.: Efficient parameter selection for support vector machines in classification and regression via model-based global optimization. In: Proc. IEEE International Joint Conference on Neural Networks, vol. 3, pp. 1431–1436 (2005)

[50] Staelin, C.: Parameter selection for support vector machines. Technical Reports, HP Labs (2002)

[51] Boardman, M., Trappenberg, T.: A heuristic for free parameter optimization with support vector machines. In: Proc. International Joint Conference on Neural Networks, pp. 610–617 (2006)

[52] Imbault, F., Lebart, K.: A stochastic optimization approach for parameter tuning of support vector machines. In: Proc. 17th International Conference on Pattern Recognition, vol. 4, pp. 597–600 (2004)

Index